HIGHER HUMAN BIOLOGY

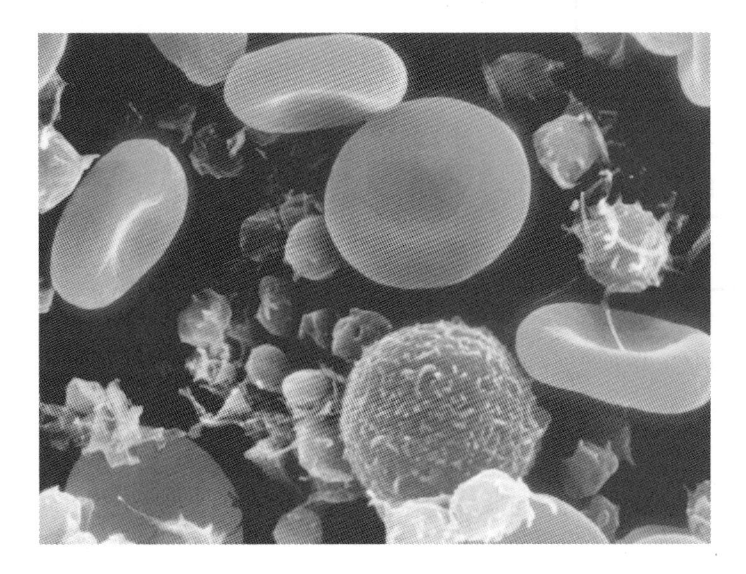

Tony Aitken, Rab Dickson
and Harry Hoey

Acknowledgements

The Publishers would like to thank the following for permission to reproduce copyright material:

Acknowledgements
Extracts from past question papers are reprinted with the permission of the Scottish Qualifications Authority. The publishers would also like to thank DART Publishing for permission to reproduce question material.

Every effort has been made to trace all copyright holders, but if any have been inadvertently overlooked the Publishers will be pleased to make the necessary arrangements at the first opportunity.

Although every effort has been made to ensure that website addresses are correct at time of going to press, Hodder Gibson cannot be held responsible for the content of any website mentioned in this book. It is sometimes possible to find a relocated web page by typing in the address of the home page for a website in the URL window of your browser.

Hachette's policy is to use papers that are natural, renewable and recyclable products and made from wood grown in sustainable forests. The logging and manufacturing processes are expected to conform to the environmental regulations of the country of origin.

Orders: please contact Bookpoint Ltd, 130 Milton Park, Abingdon, Oxon OX14 4SB. Telephone: (44) 01235 827720. Fax: (44) 01235 400454. Lines are open 9.00 – 5.00, Monday to Saturday, with a 24-hour message answering service. Visit our website at www.hoddereducation.co.uk. Hodder Gibson can be contacted direct on: Tel: 0141 848 1609; Fax: 0141 889 6315; email: hoddergibson@hodder.co.uk

First published in 2007 by
Hodder Gibson, an imprint of Hodder Education,
part of Hachette Livre UK
2a Christie Street
Paisley PA1 1NB

Impression number 5 4 3 2
Year 2010 2009 2008

Cover photo © PHOTOTAKE Inc. / Alamy
Typeset in 9.5 on 12.5pt Frutiger Light by Phoenix Photosetting, Chatham, Kent
Printed and bound in Great Britain by Martins The Printers, Berwick-upon-Tweed

A catalogue record for this title is available from the British Library

ISBN-13: 978-0-340-92705-2

CONTENTS

INTRODUCTION

This book is a personal study guide to help you pass Higher Human Biology. To achieve a suitably high grade you will have to fulfil the learning outcomes of the course and have developed skills in Problem Solving, Practical Abilities and in answering examination questions. This book has been written to help you meet all these requirements.

You might find it useful to have a copy of the Arrangements Document for Higher Human Biology as it contains the course content that is tested in the National Examinations, together with other course details. Your school may already have supplied you with this but you can also access it from the SQA website (www.sqa.org.uk).

The first chapter in the book helps you to identify the skills required for Problem Solving, Practical Abilities and answering examination questions. In the other chapters the learning outcomes of the course content have been clearly written. 'Hints and Tips' and 'For Practice' tasks are included to help reinforce your understanding of the learning outcomes. The 'Exam Questions' at the end of each chapter, together with the commentary on how to answer, will help to further reinforce and develop the skills necessary for examination success.

One of the tips given in each chapter is that you should produce 'Flash Cards' for the words and phrases that appear in bold type in the text. 'Flash Cards' are an excellent learning aid and can be used by an individual, in pairs or in a group. On one side of the card you write a biological process or structure and on the reverse side you write the definition of the process or a description of the function.

Side 1	Side 2
Metabolism	All the chemical reactions that make up all the biochemical pathways

Both sides can be used to ask a question.
Side 1: What is the definition of metabolism?
Side 2: What term is described by 'All the chemical reactions that make up all the biochemical pathways'?

Cards can be made by cutting a piece of paper into rectangular shapes, approximately 10cm × 3cm. To keep the cards for a chapter together, punch a hole in one of the corners and keep them in a key ring. Carry the key ring with you and you can study the words for that chapter at any time and anywhere.

We hope that you enjoy using this study guide and that it helps you to be better prepared for your examination.

Tony Aitken, Rab Dickson and Harry Hoey

Chapter 1

SKILLS IN ANSWERING EXAM QUESTIONS

The development, through practice, of a wide range of skills is required for success in Higher Human Biology. The more often you work at a skill the greater is your success. In this chapter you will be introduced to methods that help in answering questions based on course skills. Skills for Higher Human Biology include:

- skills for answering Multiple Choice questions
- skills for answering Extended Writing questions
- skills for answering Problem Solving questions
- skills for answering Practical Abilities questions.

Problem Solving (PS) and Practical Abilities (PA) questions make up on average 38 of the 130 available marks in the National Examination. This is approximately 30% of the available marks. You must score well in PS and PA to score a high grade pass. The examination is out of 130 marks and the time allocated is 150 minutes. This works out to give you an average time of around 1 minute 15 seconds per mark.

Skills in answering Multiple Choice questions

In the Higher Human Biology exam, Section A consists of 30 Multiple Choice (MC) questions. The Arrangements state that '*between 9 and 11 questions test Problem Solving and Practical Abilities, and the rest test Knowledge and Understanding*'.

Hints and Tips

Always read each question very carefully. Highlight or underline key words.

The sequence of questions roughly follows the order in which topics appear in the Arrangements Documents.

See if you can answer the question **without looking** at the four options. If you think that you know the answer, then look for the option that best matches.

If a question is giving you difficulty, try first to remove any of the options that you think are clearly wrong. Continue to remove options until you are left with only one. This has to be the answer, even if you don't know for certain that it is correct.

There is usually a fairly equal distribution of answers A, B, C and D. If you have time at the end of the examination it is worth checking this.

Short answer questions

In the National Examination Section B consists of structured questions. The Course Specifications for Section B are as follows:

This section will contain structured questions with an allocation of 80 marks. Between 25 and 30 marks test Problem Solving and Practical Abilities and the remainder test Knowledge and Understanding.

An indication of the number of points required for an answer can be determined by:

- the number of marks awarded for a question
- the number of lines provided in the answer space.

Easier question types

These are questions that call for straightforward recall. These questions start with words like *Name*, *State*, *Give*, *Label*, *Where*, *What*, *How* and *Complete the table*.

These ask for the facts that a hardworking candidate can memorise for at least a C pass. Frequent use of flash cards is a very useful way to memorise the basic facts.

Harder question types

Questions that include the instruction to *Explain*, *Describe*, *Suggest*, *What evidence* and *Give a reason for your answer* are almost always of a higher level of demand. These 'A type' questions are put into the exam to discover who deserves a B, or even an A. They work because candidates with bitty knowledge cannot marshal all the facts required to answer the question correctly. Frequent use of flashcards will help to fill in the spaces in the facts to allow many of this type of question to be answered.

Many of these 'thinking' questions come up often, so a hardworking candidate who works through the past papers thoroughly will be able to convert some of the thinking questions into knowledge. The more 'thinking' questions that you can convert into learned facts the better.

Hints and Tips

Focus hard on working out what the questioner is expecting.

Try to write down all the facts that you consider relevant to the question.

Marks are not deducted in Section B for additional information, unless it contradicts the correct answer.

Abbreviations are acceptable for many terms – DNA for deoxyribonucleic acid, ADH for anti-diuretic hormone. These are shown in the Arrangements Document.

Skills in answering extended response questions (essays)

Section C in the National Examination consists of extended response questions. The course specifications for Section C are as follows:

This section will consist of four extended response questions to test the candidate's ability to select, organise and present relevant knowledge. Section C will have an allocation of 20 marks and will include:

Two structured extended response questions for 10 marks. Candidates will be expected to answer one of these questions.

Two open extended response questions for 10 marks (1 mark for relevance, 1 mark for coherence and 8 marks for knowledge and understanding).

Candidates will be expected to answer one of these questions.

Section C from a National Examination is shown here. The choice in the structured extended response questions is shown as 1A and B. The marks allocated to each area are shown. The choice in the open extended response questions is shown as 2A and B.

SECTION C — *Marks*

Both questions in this section should be attempted.

Note that each question contains a choice.

Questions 1 and 2 should be attempted on the blank pages which follow. Supplementary sheets, if required, may be obtained from the invigilator. Labelled diagrams may be used where appropriate.

1. Answer **either** A **or** B.

A. Describe the functions of the liver under the following headings:

(i) production of urea; **2**

(ii) metabolism of carbohydrates; **5**

(iii) breakdown of red blood cells. **3**

(10)

OR

B. Describe the cardiac cycle under the following headings:

(i) nervous and hormonal control of heart beat; **4**

(ii) the conducting system of the heart. **6**

(10)

In question 2 ONE mark is available for coherence and ONE mark is available for relevance.

2. Answer **either** A **or** B.

A. Give an account of the transmission of a nerve impulse at a synapse. **(10)**

OR

B. Give an account of the carbon cycle and its disruption by human activities **(10)**

To achieve a Grade A or B pass you must be able to answer well in extended response questions. Decisions on which essays to answer are made easier if at the start of the examination you check the titles. Your brain will work subconciously in response to the titles and should, by the time you reach Section C, be more ready to make your choices. The order in which the essays are answered does not matter.

On the basis of the level of difficulty and time available per mark, you should allocate at least 30 minutes to Section C. The titles must be read carefully. Each question can be divided into areas. You can only gain full marks if you have attempted to answer every area within a question.

You should identify each area within a question and write these down. Tick off each area as you answer the question.

In extended response questions each piece of relevant knowledge can be seen to have a value of 1 mark. In the structured question the mark allocation for each area is an indicator as to the number of pieces of relevant knowledge that is required. In the open question there is no indication of how the 8 available marks are to be allocated.

In this case you must write down every last piece of **relevant** information that you have learned. Check that you have presented the maximum number of points possible.

- **Diagrams** can be used in extended response questions and often help clarify your understanding of the facts. For example, a diagram representing the fluid mosaic structure of the plasma membrane can help display your knowledge of the chemicals present, their molecular arrangement and indeed some of the functions, such as channel proteins. All diagrams **must be fully labelled**.
- **Flowcharts** can be used in extended response questions. Typical questions that suit this style would include respiration and regulating mechanisms. The biggest danger is forgetting to show the direction of flow. The direction of flow **must** be shown by **arrows** – otherwise marks are lost.
- **Bullet points** can be used in extended response questions. But each bullet point must contain a clearly understood and relevant statement written as a sentence. Suitable headings **must** be used to separate each group of bullet points.

Practise the extended response type of questions. After studying a topic, ask your teacher for a suitable essay title. Study the topic area and then try to answer without looking at your notes. Check your attempt against appropriate marking instructions.

Note the areas in which you failed to score marks. Slowly but surely you will start to improve and as you improve you will gain in confidence. Section C is the most difficult area within the examination. You must train for success in this area. If you have not attempted at least 20 extended response questions by the date of the National Examination then you have not practised enough.

Problem Solving skills

The National Course Specifications that relate to Problem Solving (PS) are as follows.

1. *Select relevant information from text, tables, charts, graphs and/or diagrams.*
2. *Present information appropriately in a variety of forms, including written summaries, extended writing, tables and/or graphs.*
3. *Process information accurately using calculations where appropriate. Calculations to include percentages, averages and/or ratios. Significant figures and units should be used appropriately.*

PS Skills have to be developed throughout the coursework. In the National Examination these skills have to be applied within an unfamiliar situation. Don't be put off PS because you think that you are not very good at maths. Many of the skills can be mastered with practice.

The calculations are normally straightforward. Numerical answers in PS will usually be a whole number or worked out to one decimal point at the most.

Hints and Tips

Always investigate the scales of the graphs and calculate the values of the smallest divisions shown in the graphs for both the X and Y axes.

Use a ruler to read from the graph to the values on the X and Y axes.

If you get a complex answer to the calculation, check the numbers that you used in the calculation and check by recalculating.

If there are two curves within the one graph, or two Y axes in the graph, then highlight the curves or axes to make sure that you read from the correct one.

Practise data handling questions from past papers.

Use questions from Higher Biology as well as Higher Human Biology. The skills required are the same and are independent of knowledge of the course content.

Skills in Practical Abilities

For Higher Level, in the National External Examination, the assessment of Practical Abilities (PA) is based on generic skills. These are skills that are developed throughout the experimental coursework and that are then applied within an unfamiliar experimental situation. Remember that in the National Examination the experimental situation question will be based on an experiment that you may not have carried out during the coursework.

The National Course Specifications that relate to PA are as follows:

4. *Plan and design experimental procedures to test given hypotheses or to illustrate particular effects. This could include identification of variables, controls and measurements or observations required.*
5. *Evaluate experimental procedures in situations that are unfamiliar, by commenting on the purpose of approach, the suitability and effectiveness of procedures, the control of variables, the limitations of equipment, possible sources of error and/or suggestions for improvement.*
6. *Draw valid conclusions and give explanations supported by evidence. Conclusions should include reference to the overall pattern to readings or observations, trends in results or comment on the connection between variables and controls.*
7. *Make predictions and generalisations based on available evidence.*

Some terms commonly used in experimental situation questions include:

validity of procedure
accuracy of readings
reliability of results.

Validity is to do with the 'correctness' or 'fairness' of the experimental procedure.

The basic idea is that only one factor should be varied within the experimental procedure and that all the other factors (variables) that may affect the results obtained must be kept the same.

Accuracy refers to the precision or exactness of the measurements recorded during the experiment. For example, a basic thermometer will be accurate to plus or minus 1°C.

Reliability is to do with the 'believability' of the results obtained when the procedure used is valid and the accuracy of the results is acceptable.

Hints and Tips

Graphs do not have to start at 0.

For Higher Human Biology do **not** try to do a line of best fit.

In questions that ask for 'variables not already mentioned', check through the data and highlight any variable that is mentioned.

The only way to allow a valid comparison if masses/lengths/etc. are not the same value at the start of the experiment is to use 'percentage change'.

When describing a trend and/or pattern in the results, you must identify the values asked for at the appropriate points that changes take place.

Precautions to minimise errors would include:

- wash out a piece of apparatus, e.g. a syringe, to prevent cross-contamination
- ensure that procedures were identical in all experiments with the exception of the controlled variable.

Chapter 2

THE ROLE OF ENZYMES IN CELL METABOLISM

Catalytic activity

Key Points

A **catalyst** speeds up the rate of a chemical reaction by lowering the energy needed to make the reaction take place. The catalyst remains unchanged after the reaction and can carry out the same reaction many times.

Enzymes are proteins that act as catalysts in **biochemical pathways**. Enzymes can act within cells (in respiration), and outwith cells (in digestion of food).

Metabolism refers to all the chemical reactions that make up all the biochemical pathways in an organism. Each stage in a pathway is controlled by an enzyme.

Figure 2.1 outlines a biochemical pathway within a cell in which enzyme 1 converts compound A to compound B and enzyme 2 converts compound B to compound C.

Figure 2.1 Biochemical pathway within a cell

Inborn or genetically inherited metabolic errors involve the absence of enzymes. In Figure 2.1, if enzyme 1 was not synthesised (was absent) then compound A would not be converted to compound B. The biochemical pathway would not be completed as no compound B would be present to be converted to compound C.

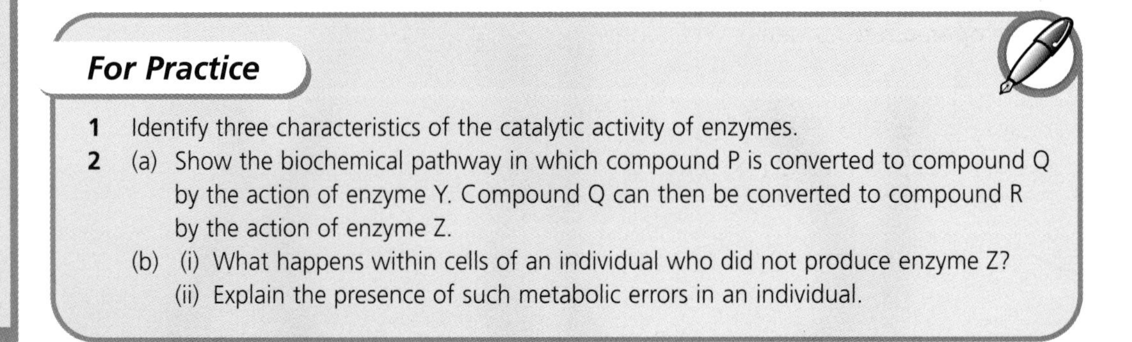

For Practice

1 Identify three characteristics of the catalytic activity of enzymes.

2 (a) Show the biochemical pathway in which compound P is converted to compound Q by the action of enzyme Y. Compound Q can then be converted to compound R by the action of enzyme Z.

(b) (i) What happens within cells of an individual who did not produce enzyme Z?

(ii) Explain the presence of such metabolic errors in an individual.

An outline of enzyme activity is shown in Figure 2.2.

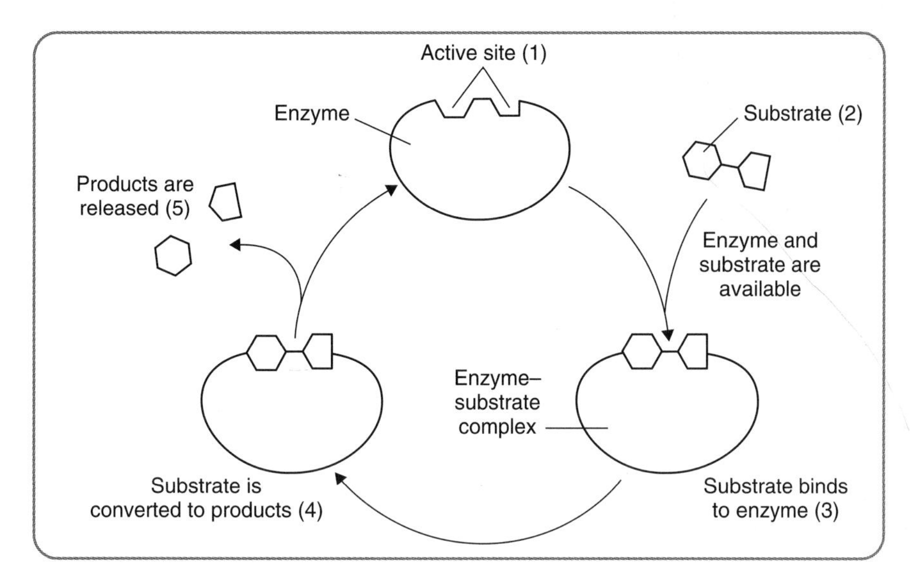

Figure 2.2 A reaction catalysed by an enzyme

The **active site** (1) is the area of the enzyme in which the catalysed reaction takes place.

The chemical that the enzyme reacts with is the **substrate** (2).

Substrate molecules bind to the enzyme at the active site (3).

Substrate is converted to **products** (4).

Products are released from the enzyme (5).

The active site is available for another substrate molecule (1).

Enzymes are substrate specific as only one substrate has the correct shape to bind to the active site.

Factors that affect enzyme activity include temperature, pH, enzyme concentration, substrate concentration and the presence of inhibitors.

Figure 2.3a shows the effect of temperature on enzyme activity. As temperature increases, the reaction rate of the enzyme increases. The rate is fastest at 37°C – this is the **optimum** temperature. With further increase in temperature beyond the optimum the reaction rate decreases and finally stops. This is because enzymes are proteins and proteins are **denatured** at high temperatures. The shape of the active site is changed.

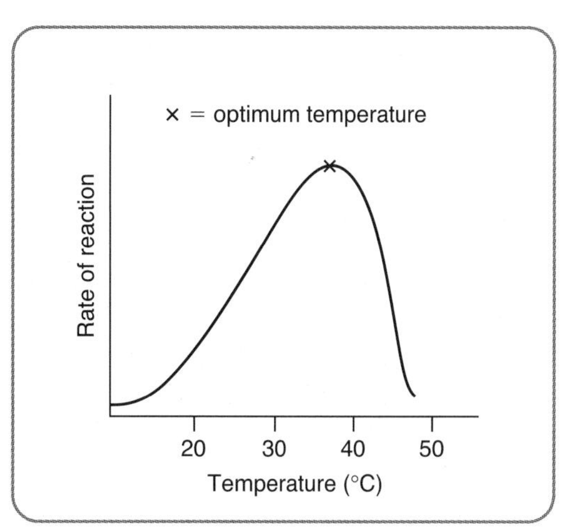

Figure 2.3a

Figure 2.3b shows the effect of pH on enzyme activity. Enzymes are active over a range of pH. The pH at which the reaction rate is fastest is the **optimum** pH. The stomach is the only area in the body that is highly acid. The protein-digesting enzyme pepsin, present in the stomach, has an optimum of pH 2.

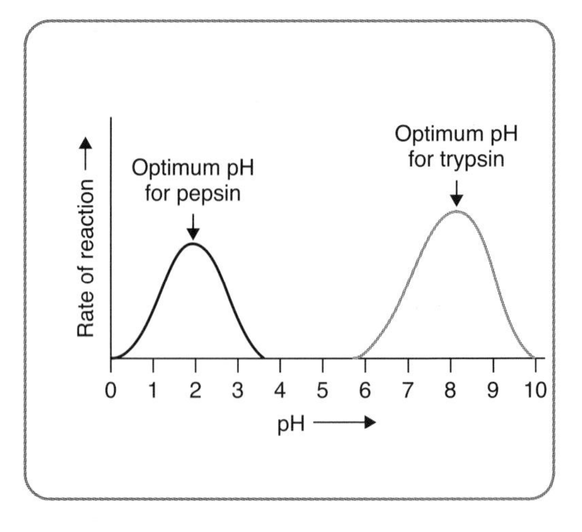

Figure 2.3b

Figure 2.3c shows the effect of increasing enzyme concentration on the reaction rate of an enzyme when availability of substrate is unlimited. As enzyme concentration increases the reaction rate increases. With more enzyme present there are more active sites for substrate molecules to bind to, so substrate molecules are broken down faster.

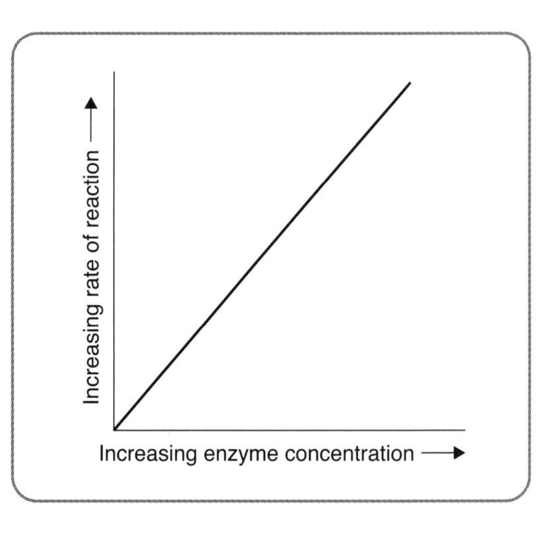

Figure 2.3c

Figure 2.3d shows the effect of increasing substrate concentration on the reaction rate of an enzyme. At low concentrations there are not enough substrate molecules to bind to all the active sites that are available. As substrate concentration increases, more active sites will be filled and the reaction rate increases. The reaction rate continues to increase until at a certain substrate concentration all the active sites will be working at their maximum. Any further increase in substrate concentration cannot increase the reaction rate. This is shown by the graph levelling off.

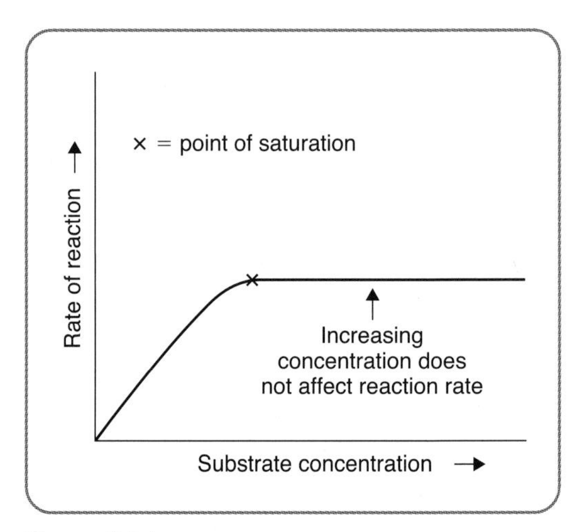

Figure 2.3d

Enzyme inhibition is due to the presence of chemicals other than the substrate. These **inhibitors** lead to a decrease in the reaction rate of the enzyme. Two types of inhibition are **competitive** and **noncompetitive** inhibition. From Figure 2.4 it can be seen that competitive inhibition occurs when a chemical with a structure similar to the substrate is present. Molecules of this chemical can bind to the active site of the enzyme. Both the substrate molecules and the inhibitor molecules are in competition for the active site of the enzymes. 'Success' in the competition varies with the concentration of both inhibitor and substrate. Competitive inhibition reduces the reaction rate of the enzyme at low substrate concentrations.

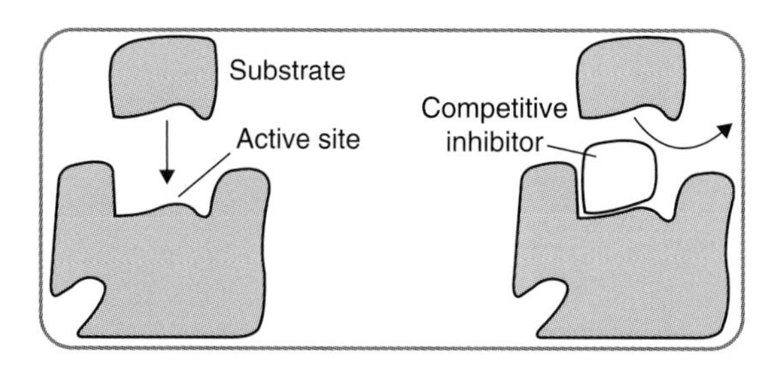

Figure 2.4 Competitive inhibition

Figure 2.5 shows how a **noncompetitive inhibitor** binds to the enzyme. A noncompetitive inhibitor molecule binds to an area other than the active site and this causes the shape of the active site to be changed. Substrate molecules can no longer bind to the altered shape at the active site. The reaction rate decreases and remains low even at high substrate concentrations because the shape of the active site remains altered in affected enzyme molecules.

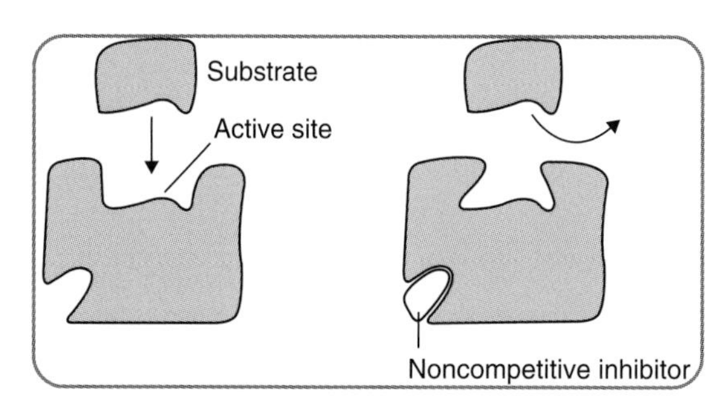

Figure 2.5 Noncompetitive inhibition

Many enzymes are produced in an inactive form. **Activators** have to be present to convert the enzyme from an inactive form to the active form. Types of activator include mineral ions, vitamins and enzymes. Mineral ions and vitamins that are required as activators are called **co-enzymes**. Figure 2.6 shows how a co-enzyme is involved in activation of an enzyme. The co-enzyme and substrate molecules bond to form a co-enzyme substrate complex. The complex binds to the active site of the enzyme.

Many digestive enzymes are secreted into the cavity of the gut in an inactive form. Other enzymes present in the cavity of the gut cut away a part of the inactive form of the enzyme to produce the active form.

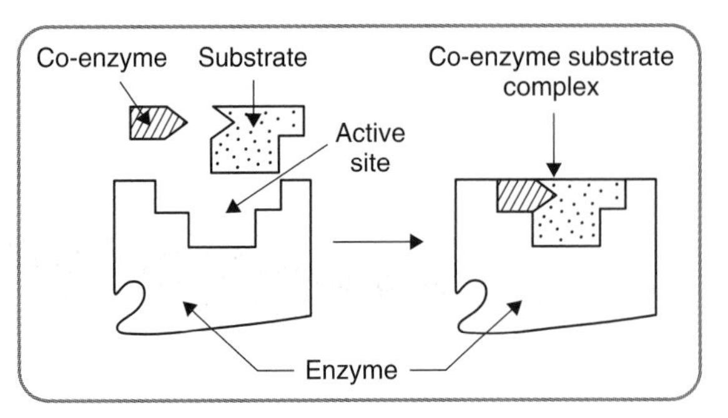

Figure 2.6 Activation of an enzyme

The equations below show how enzymes act as enzyme activators.

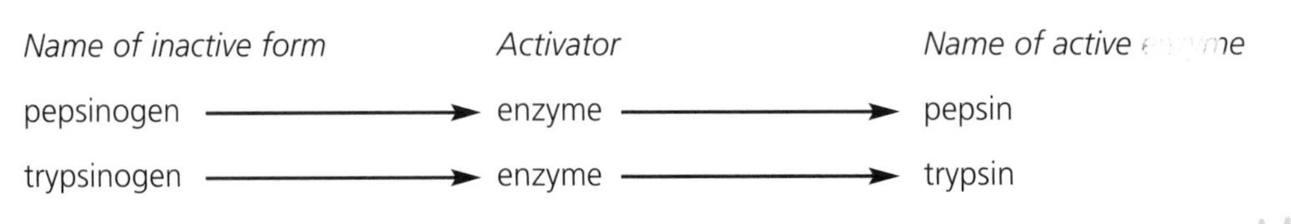

Name of inactive form		*Activator*		*Name of active enzyme*
pepsinogen	→	enzyme	→	pepsin
trypsinogen	→	enzyme	→	trypsin

Hints and Tips

Later in the course you will see that the mutation involved in the disorder phenylketonuria (PKU) is a genetically inherited enzyme disorder.

In graphs that relate to competitive inhibition, the reaction rate is lower than with no inhibitor at low substrate concentrations and is the same at high substrate concentrations.

In graphs that relate to noncompetitive inhibition, the reaction rate remains much lower than that of the reaction rate when no inhibitor is present.

In diagrams, if a molecule binds to the active site this is competitive inhibition.

In diagrams, if a molecule binds to an area other than the active site this is noncompetitive inhibition.

Enzymes are produced in an inactive form to prevent self-digestion of the cell contents.

For Practice

1 Copy and complete the table below to show details of five factors that affect enzyme activity. Temperature is shown for you.

Factor	Details of how factor affects enzyme activity
Temperature	As temperature increases reaction rate increases to optimum at 37°C. At higher temperatures the reaction rate decreases and finally stops as the enzyme is denatured.

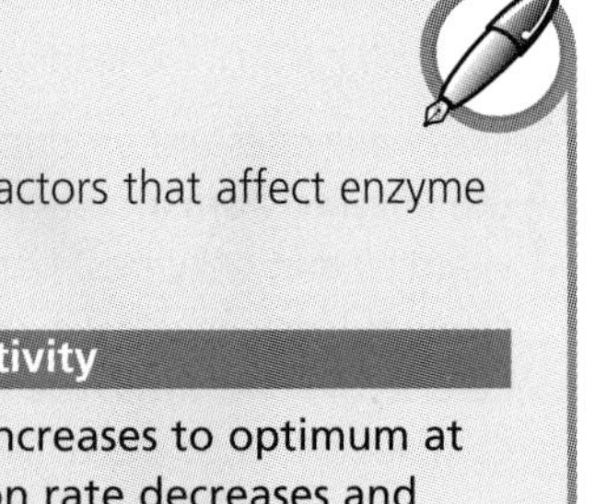

For Practice *continued* ➢

For Practice continued

2 Copy and complete the table below to show similarities and differences between competitive and noncompetitive inhibition.

Competitive inhibition	Noncompetitive inhibition
1 Reduced reaction rate	1
2	2 Inhibitor molecule does not compete for the active site
3 Shape of active site is unaltered	3
4 Inhibitor molecule binds to active site	4
5	5 Reaction rate remains low at high substrate concentrations

3 Copy and complete these sentences which relate to activation of enzymes.
Three types of enzyme activators are_____________, vitamins and _________.
Mineral ions and _____________ are called _________________. Co-enzyme molecules bond with _________________ molecules. Enzymes act as an _______________ by converting an _____________ form of a digestive enzyme such as pepsinogen to an _________ form called ____________.

4 Make flashcards for all of the words that appear in **bold** type in the text of this chapter.

Exam Questions

1 The specificity of an enzyme is related to the:

A pH and temperature
B shape of the active site
C presence of a competitive inhibitor
D substrate concentration.

2 A biochemical pathway is shown below.

$$P \xrightarrow{\textit{enzyme 1}} Q \xrightarrow{\textit{enzyme 2}} R \xrightarrow{\textit{enzyme 3}} S$$

A noncompetitive inhibitor that affects enzyme 2 is introduced into the pathway.

Which concentration change occurs?

A Increased in both compounds Q and R
B Increased in compound Q and decreased in R
C Decreased in both compounds P and Q
D Decreased in compound Q and increased in P

Exam Questions continued

3 Figure 2.7 represents a reaction that is catalysed by an enzyme.

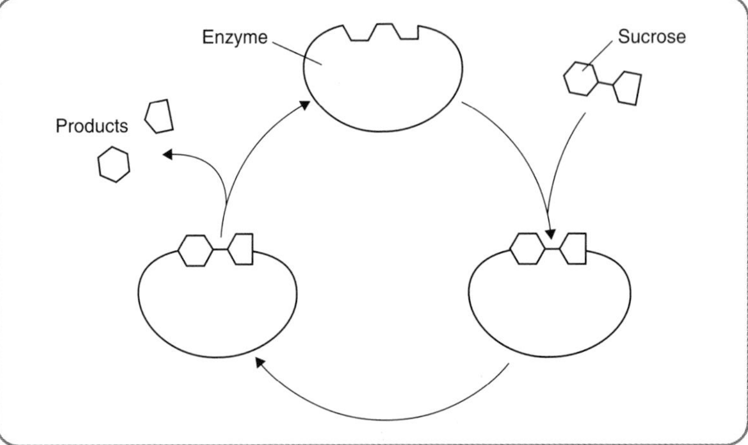
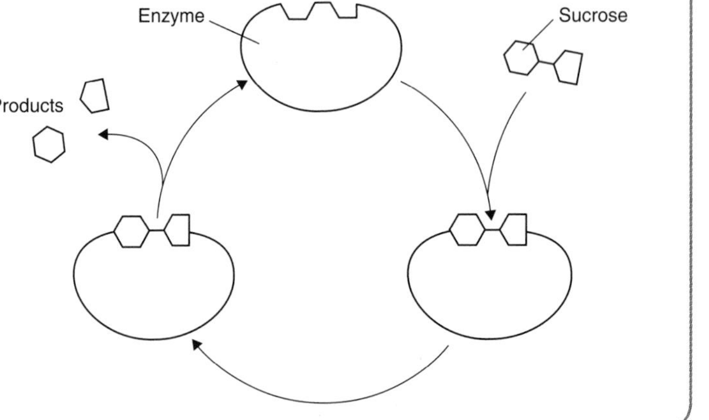

Figure 2.7 A reaction catalysed by an enzyme

(a) Name the area of the enzyme where the reaction takes place. *(1)*

(b) Cyanide is a noncompetitive inhibitor for this enzyme. Suggest how cyanide inhibits the enzyme. *(2)*

(c) Explain why some enzyme-catalysed reactions require the presence of vitamins. *(1)*

(d) Figure 2.8 shows the effect of substrate concentration on the reaction rate of this enzyme when the pH and temperature are set to their optimal values.

(i) Explain why the graph levels out between points X and Y. *(1)*

(ii) How could the reaction rate be made to increase between points X and Y? *(1)*

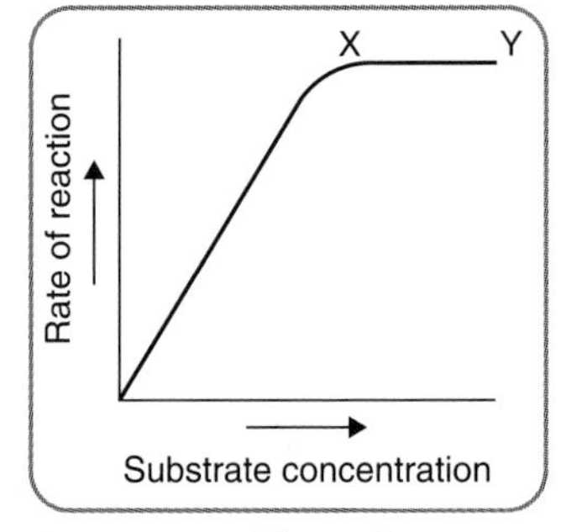

Figure 2.8 Effect of substrate concentration on the reaction rate of an enzyme

4 Figure 2.9 shows the effects of the presence of a competitive and a noncompetitive inhibitor on the rate of an enzyme-catalysed reaction.

(a) Which line in Figure 2.9 shows the change in the reaction rate of the enzyme when in the presence of the noncompetitive inhibitor? Justify your choice. *(2)*

(b) Explain the meaning of competitive inhibition. *(2)*

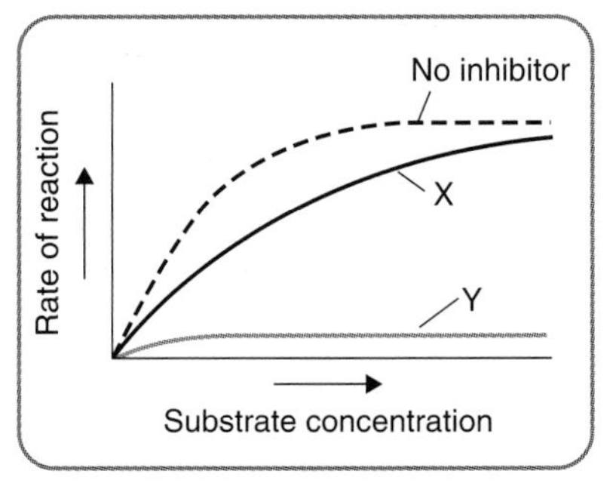

Figure 2.9 Effects of inhibitors on an enzyme-catalysed reaction

5 Give an account of enzymes under the following headings:

(a) Catalytic activity in cell metabolism *(6)*

(b) Competitive inhibition. *(4)*

Answers

1 You should know that specificity relates to the substrate and that due to its shape it can bind to the active site. **Answer** B

2 You should know that if a noncompetitive inhibitor affects enzyme 2 then less of compound Q is converted to R. Compound Q will increase as less is converted and R will decrease as it will still be converted to S. **Answer** B

3 (a) You should know that the reaction takes place in the active site.

(b) In this answer you have to state how a noncompetitive inhibitor attaches to the enzyme and how this stops the enzyme from functioning. Cyanide attaches to an area of the enzyme other than the active site (1st mark). The shape of the active site is altered and the substrate can no longer bind (2nd mark).

(c) You should know that vitamins are co-enzymes that bond to substrate molecules.

Answer The substrate can only fit to the active site if it is attached to a vitamin.

(d) (i) You should know that the graph levels out when all the active sites are working at their maximum.

Answer The substrate concentration is high between X and Y and all active sites are filled and are working at their maximum.

(ii) You are told that temperature and pH are at optimum and, from the graph, an increase in substrate concentration does not increase the reaction rate. The only other factor that would increase the reaction rate must be the enzyme.

Answer By addition of more enzyme.

4 (a) You should know that a noncompetitive inhibitor makes the active site no longer available to the substrate.

Answer Line Y.

Justification: An increase in substrate concentration did not increase the reaction rate. This matches the fact that noncompetitive inhibitors cause active sites to be permanently unavailable to the substrate.

(b) You should know that competitive inhibitors compete with the substrate for the active sites and this is related to the structure of the molecules.

Answer A competitive inhibitor molecule has a similar structure to the substrate (1st mark). Competes with the substrate molecule for the active site (2nd mark).

Answers continued

5

Nos	Mark scoring points		Comments
	(a) Catalytic activity in cell metabolism		*Write the sub-heading*
1	Catalysts speed up the rate of a chemical reaction	1	*Give as much detail as you can on catalysts*
2	Catalysts lower the energy required to make the reaction take place	1	
3	Catalysts remain unchanged after the reaction	1	
4	Catalysts can carry out the same reaction many times	1	
5	Metabolism is all the chemical reactions within an organism	1	*Give a definition of metabolism*
6	Enzymes catalyse reactions within and outside cells	1	*Give detail of reactions within and outside the cell*
7	Respiration as an example within cells *or* digestion of food for outside cells	1	
8	Genetically inherited errors of metabolism involve absence of / incorrectly synthesised enzymes	1	*PKU as an example of an error in cell metabolism*
			*Any points to a maximum of **6 marks***
	(b) Competitive inhibition		*Write the sub-heading*
10	Competitive inhibitor molecules are similar in structure to those of substrates		*Describe competitive inhibition in terms of substrate molecules fitting to the active site*
11	Molecules of competitive inhibitor can also bind to the active site		
12	Substrate molecules and competitive inhibitor molecules compete for the active sites of the enzymes		
13	Competitive inhibition reduces the reaction rate of the enzyme		*Describe effects of competitive inhibition*
14	Success in competition varies with concentration of inhibitor / substrate		
			*Any points to a maximum of **4 marks***

Chapter 3

PROTEIN SYNTHESIS

Protein structure

The sub-units of proteins are amino acids. In each type of protein, amino acids are linked by **peptide bonds** in a specific sequence to form a chain. The length of the chain varies in each type of protein. Figure 3.1 shows the synthesis of part of a protein chain through the linking of amino acids by peptide bonds. The chain formed shows the **primary structure** of a protein.

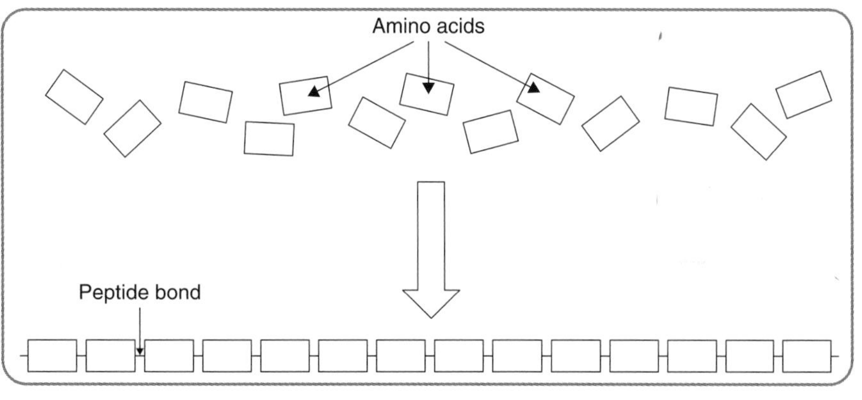

Figure 3.1 Formation of peptide bonds

Hydrogen bonds form between amino acids along the length of the chain. These hydrogen bonds alter the shape of the chain. Figure 3.2 shows how the formation of hydrogen bonds alters the shape of the chain to produce either a helical or pleated structure. The changed shape due to formation of hydrogen bonds is referred to as the **secondary structure** of the protein.

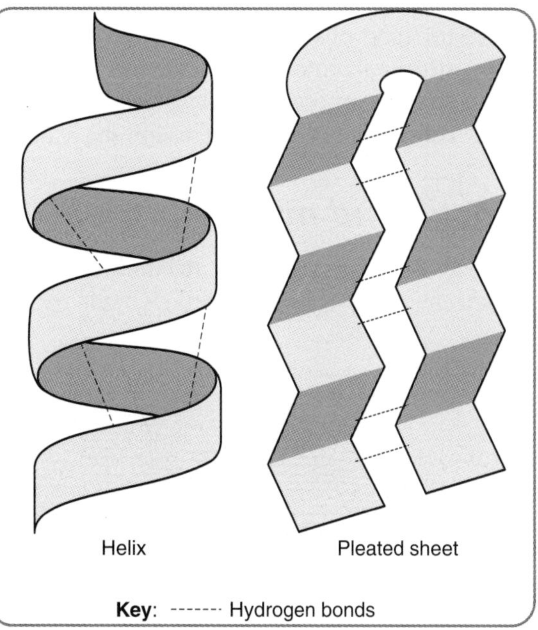

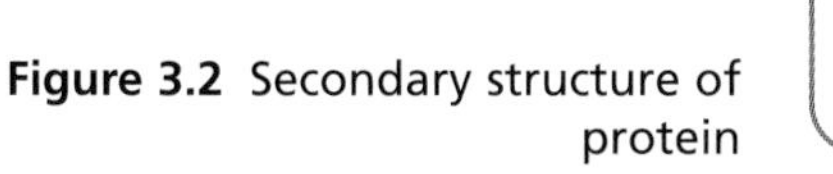

Figure 3.2 Secondary structure of protein

Additional bonding, including hydrogen bonds, takes place between amino acids and this further alters the shape of the chain. Figure 3.3 shows how this bonding gives a protein a distinctive three-dimensional shape. The changed shape due to additional bonding is referred to as the **tertiary structure** of the protein.

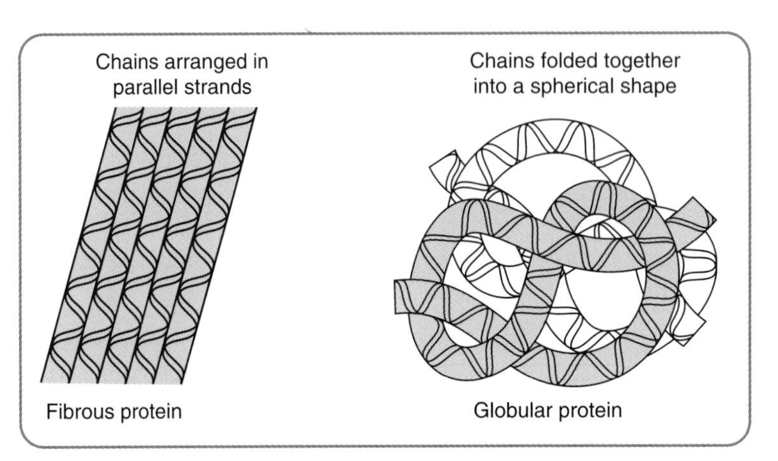

Figure 3.3 Tertiary structure of protein

Functions of proteins

Functions of proteins include:

- increasing the rate of reactions, as enzymes in biochemical pathways
- regulating blood sugar concentration, as the hormones insulin and glucagon
- controlling of growth and development, as hormones such as growth hormone
- transporting substances from one part of the body to another, for example the transport of oxygen by haemoglobin from the lungs to other parts of the body
- structure of cell membranes including membrane proteins which are involved in the transport of molecules across the cell membrane
- structural proteins giving support – these include collagen which forms part of the structure of bones, tendons and ligaments
- contraction of muscles through the action of the proteins actin and myosin.

Actin and myosin proteins

Figure 3.4 shows magnified **striated muscle**. Striations (stripes) occur due to the presence of actin and myosin proteins. Actin is made up of thin filaments and myosin of thick filaments.

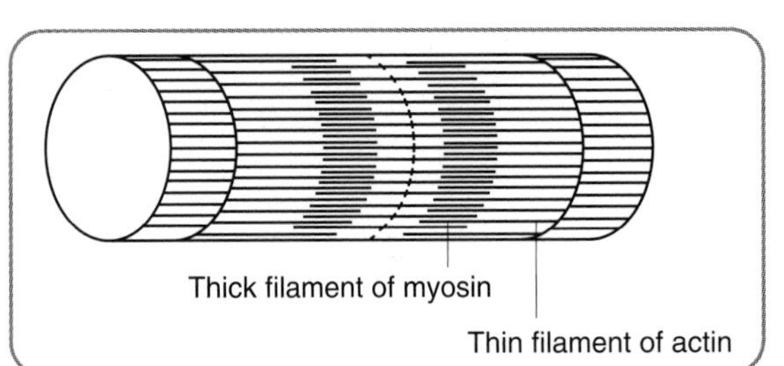

Figure 3.4 Striated muscle

Figure 3.5 shows the arrangement of the actin and myosin filaments in a relaxed muscle. A **dark band** appears in the areas where the actin and myosin filaments overlap and where myosin only is present. A **light band** appears where actin filaments only are present.

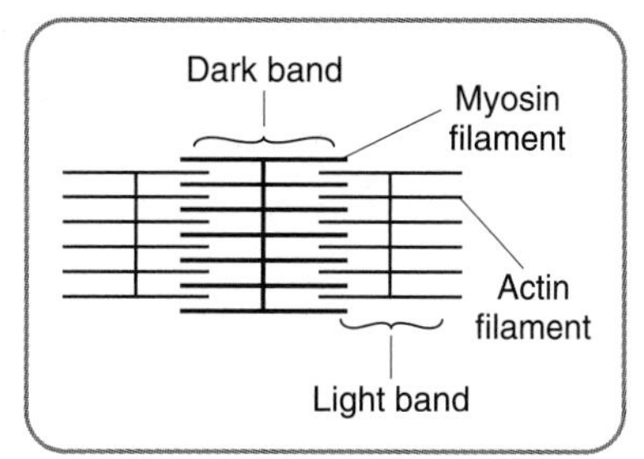

Figure 3.5 Dark and light bands of muscle

Figure 3.6 shows the arrangement of actin and myosin filaments in a contracted muscle. During contraction the filaments of actin and myosin slide in between one another – the **sliding filament theory**. During contraction the dark bands stay the same width but the light bands become narrower.

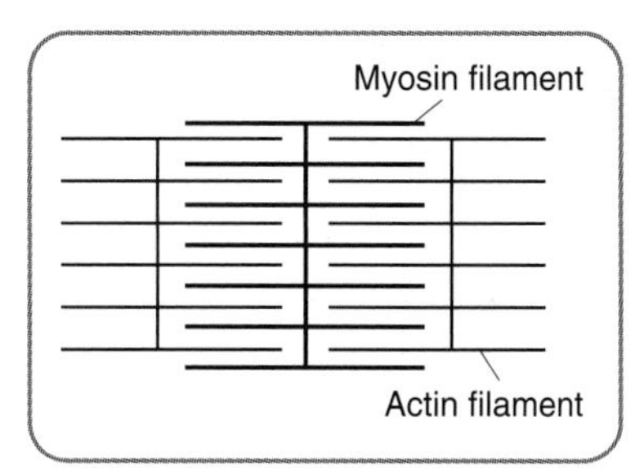

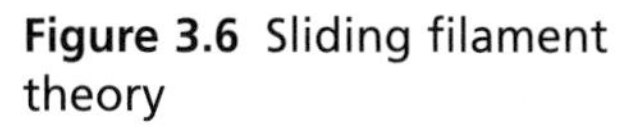

Figure 3.6 Sliding filament theory

Hints and Tips

All proteins / amino acids contain the chemical elements carbon, hydrogen, oxygen and nitrogen.

The actin and myosin proteins do not shorten. The actin and myosin filaments slide in between each other.

For Practice

1 Copy and complete the blanks in these sentences on protein structure.

The sub-units of proteins are amino acids that are linked by peptides bonds. The secondary structure is formed by hydrogen bonds. Further bonding that includes ____________ bonding causes further changes in ___________ that give rise to the ____________ structure of the protein.

For Practice continued ➢

For Practice continued

2 Copy and complete the table below on the function of types of protein, including examples. Enzymes has been completed for you.

Type of protein	Function	Example
Enzymes	To increase the rate of biochemical reactions	Glycolysis and Krebs cycle in respiration
Hormones		

Role of DNA, RNA and cellular organelles in protein synthesis

The genetic material that codes for the synthesis of proteins is present in the chromosomes of a cell. The genetic material is made of the chemical **deoxyribonucleic acid** (DNA). Certain lengths of the DNA carry the code for proteins. These areas of DNA are called genes.

Structure of deoxyribonucleic acid (DNA)

DNA is made up from basic units called **nucleotides.** Figure 3.7 shows that a nucleotide is made up of a base, the sugar deoxyribose and a phosphate.

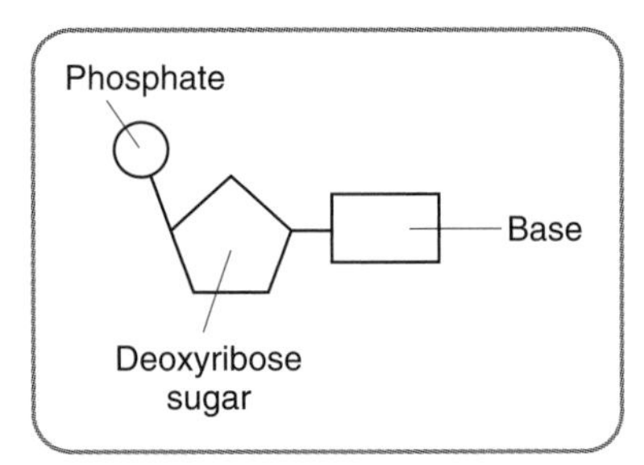

Figure 3.7 DNA nucleotide

There are four different types of nucleotide in DNA. The nucleotides have one of four different bases: adenine (A), thymine (T), guanine (G) and cytosine (C). Nucleotides bond through the phosphate group of one nucleotide and the deoxyribose sugar of an adjacent nucleotide to form a single strand (Figure 3.8). The sugar–phosphate linkages in the strands are often referred to as the **sugar–phosphate backbone**.

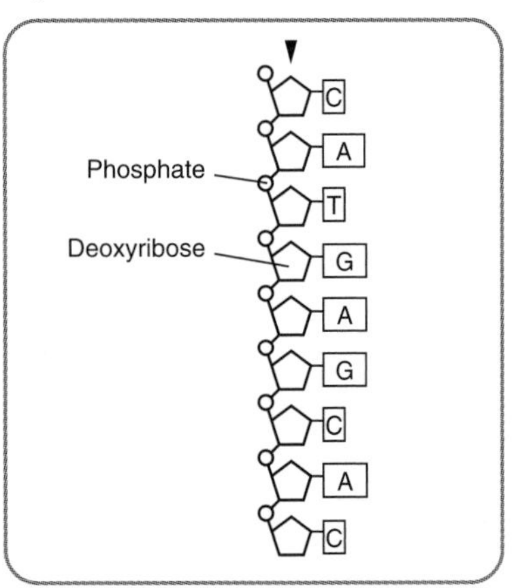

Figure 3.8 Single strand of DNA nucleotides

DNA is made up from two such strands that bond through their bases by **hydrogen bonds**. Each base can bond with one type only of the other bases. A bonds only with T, and G bonds only with C. This is **base pairing**. The base pairs are described as being complementary base pairs. Figure 3.9 shows the structure of a DNA molecule. The two strands coil to form a **double helix**. The double helix is shown in Figure 3.10.

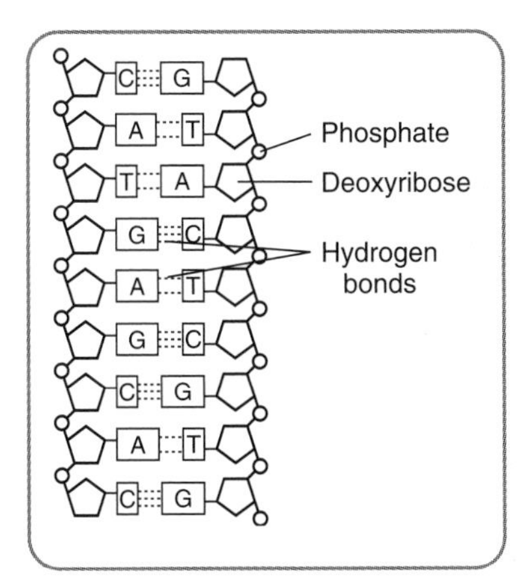

Figure 3.9 DNA molecule

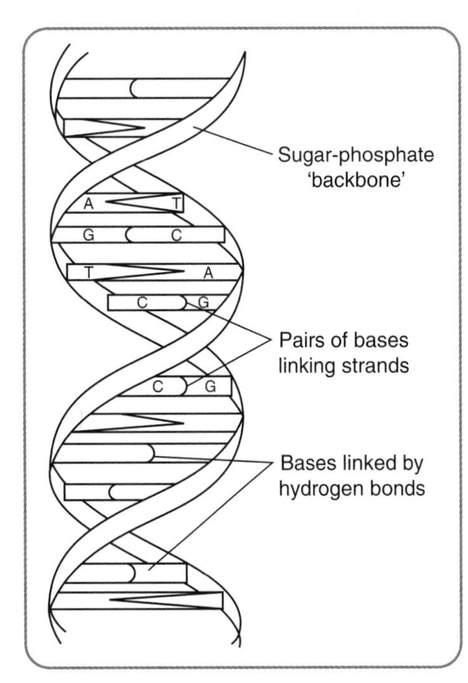

Figure 3.10 Double helix

For Practice

1 Copy and complete the blanks in these sentences on DNA structure.

The basic structural units of DNA are Nucleotides. Each one is made up of a phosphate, a sugar called Deoxyribose and a Base. DNA is composed of Single strands. Within each DNA strand the sugar of one ________ is bonded to the ________ of an adjacent nucleotide. The two strands are held together through their ________ by ________ bonds. The bases show base ________ in that the base adenine always bonds with ________ and guanine always bonds with Cytosine. The two strands coil to form a structure called a Double Helix.

2 Copy Figure 3.8 and complete the structure of the complementary strand. Label a hydrogen bond on your diagram.

Structure of RNA

RNA is made of a single strand of nucleotides. Figure 3.11 shows that an **RNA nucleotide** contains a base, the sugar ribose and a phosphate. There are four different types of RNA nucleotide. They have one of four different bases: adenine (A), **uracil** (U), guanine (G) and cytosine (C).

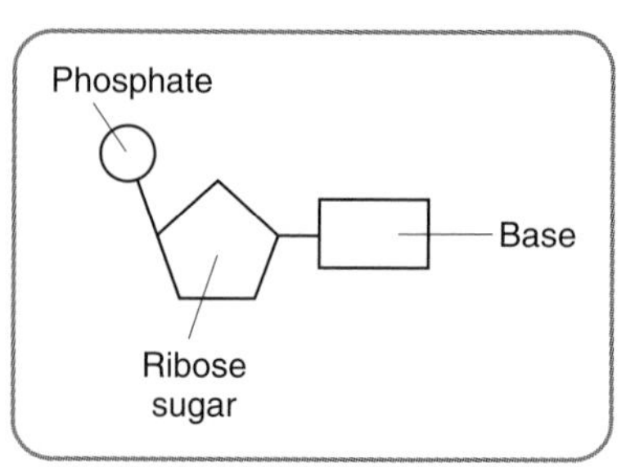

Figure 3.11 RNA nucleotide

Synthesis of RNA from DNA – transcription

The genetic code for protein synthesis is carried in the sequence of bases along the length of one of the strands of DNA. Protein synthesis is known to take place outside the nucleus on structures named ribosomes. Chromosomes, containing the DNA, are too large to pass out from the nucleus to ribosomes. To overcome this, the genetic code on the DNA is used to synthesise another molecule. The molecule that carries the code is named messenger RNA (mRNA). This is transcription – the synthesis of mRNA from DNA. The code within the sequence of bases on DNA is now in the sequence of bases of the mRNA.

Figure 3.12 outlines the synthesis of RNA. In the presence of suitable enzymes and adenosine tri-phosphate (ATP), the double helix unwinds and the hydrogen bonds holding

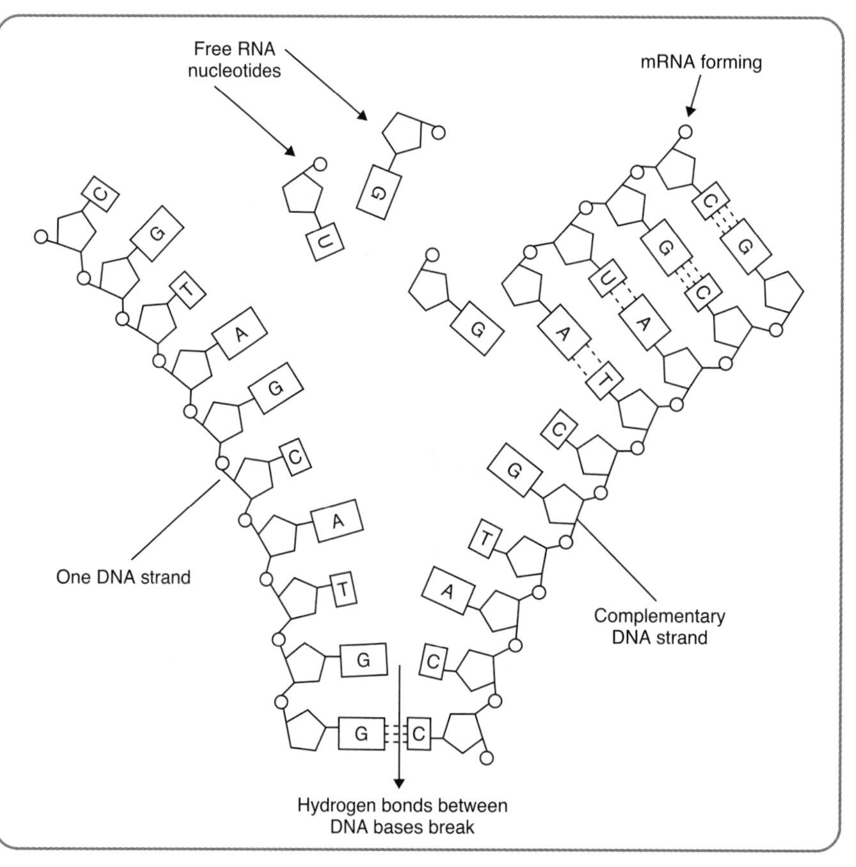

Figure 3.12 Synthesis of RNA

base pairs together in the DNA are broken. The DNA has unzipped and the bases of the DNA strands are exposed. Free RNA nucleotides attach to complementary bases on one of the DNA strands. The base pairing is A of an RNA nucleotide to a T of DNA; U of an RNA nucleotide to A of DNA; G and C of RNA nucleotides to C and G of DNA nucleotides respectively. RNA nucleotides bond through the phosphate group of one nucleotide and the ribose sugar of the adjacent nucleotide to form a single strand of mRNA.

Assembly of protein – translation

The code for the amino acid sequence in a protein is present in the sequence of bases on the **mRNA** molecule. The code is a **triplet code** and each triplet of bases on an mRNA molecule is a **codon**. There are specific codons for each of the 20 different amino acids. The mRNA breaks away from the DNA strand and passes through a nuclear pore and into the cytoplasm where it attaches to a ribosome. Amino acids are transported to the ribosome by transfer RNA **(tRNA)** molecules. Figure 3.13 shows the structure of a tRNA molecule. A tRNA molecule consists of a single strand that folds back to form a cloverleaf shape. At one end of the tRNA is a triplet of bases that form the **anticodon.** The bases of the anticodon are complementary to a codon on the mRNA. At the other end is the attachment site for a specific amino acid.

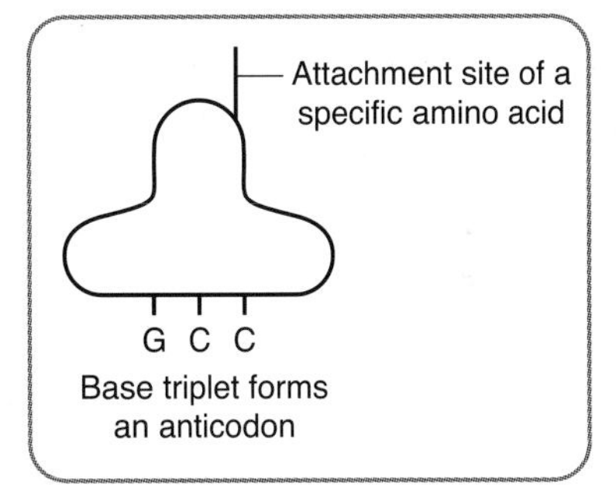

Figure 3.13 Structure of a tRNA molecule

Follow the pathway in Figure 3.14 which shows the stages of protein synthesis.

Stage 1 A tRNA transports its specific amino acid to the ribosome.

Stage 2 The anticodon of the tRNA bonds to its complementary codon on the mRNA (anticodon CCA matches to codon GGU).

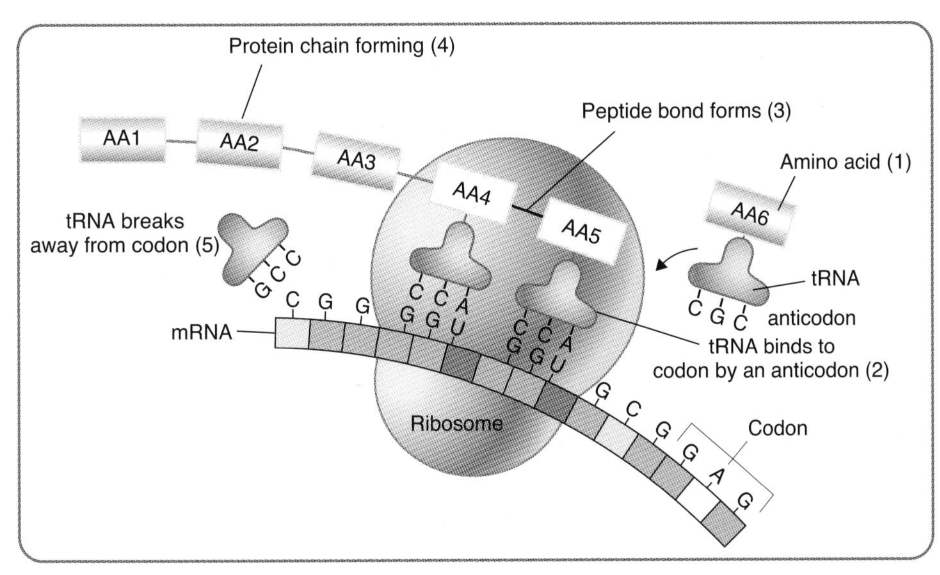

Figure 3.14 Stages of protein synthesis

Stage 3 The amino acids on adjacent tRNAs are bonded by a peptide bond.

Stage 4 The protein chain is forming as more tRNAs match to their corresponding codons and the amino acids are bonded.

Stage 5 The tRNA breaks away from the mRNA after its amino acid is bonded. This process continues until all the codons have been 'read' and the protein is formed.

Structure and role of the nucleus and nucleolus in RNA synthesis and transport

The nucleolus is present within the nucleus (Figure 3.15). The **nucleolus** is the site of synthesis of ribosomal RNA and the sub-units that make up ribosomes. The sub-units and ribosomal RNA pass out through pores in the nuclear membrane and are assembled into ribosomes. Ribosomes occur freely in the cytoplasm or are attached to the rough endoplasmic reticulum (see Chapter 5 Cell transport). **Ribosomes** are the site of translation of mRNA into protein.

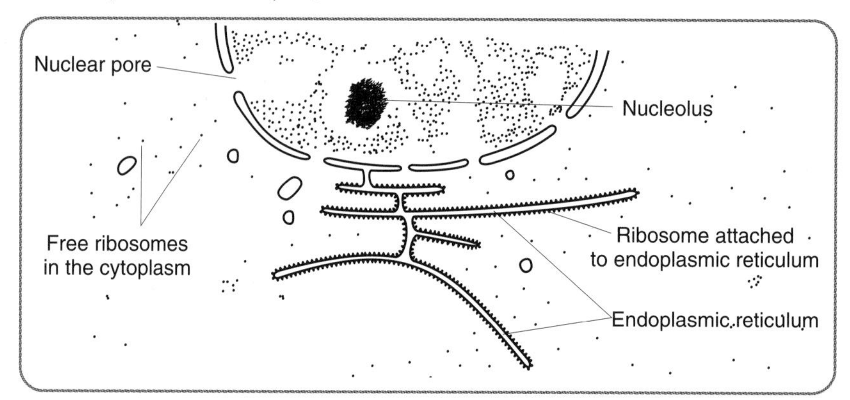

Figure 3.15 Nucleolus in nucleus

Hints and Tips

Transcribed means to 'copy across'. The code of the DNA is 'copied across' to the mRNA.

Translation means to convert words from one language to another. The 'language' of the base sequence is converted to the 'language' of the amino acid sequence.

Remember that T is present in DNA only and U in RNA only.

Polypeptide is another term used in the description of protein.

Be able to convert the code of DNA into codons of mRNA and then into anticodons of tRNA, and vice versa: ACT ↔ UGA ↔ ACU.

Use the term *complementary* to describe matching base pairs and matching codons and anticodons.

Use the term *specific* to describe the fact that each tRNA type transports one type only of amino acid.

Try to visualise the processes of transcription and translation as if it was an animation that you were watching.

For Practice

1 Copy and complete the table below to show the differences in structure of DNA and RNA molecules.

Structure	DNA	RNA
Sugar present		
Number of strands		
Bases present		

2 Copy and complete these sentences on transcription.

In the presence of suitable enzymes and _______ the DNA unwinds and the _______ bonds holding the bases together are broken. Free ______ nucleotides attach to their __________ bases on one of the DNA _______.

The base pairings are A of RNA to __ of DNA, __ of RNA to A of DNA, G of RNA to __ of DNA and __ of RNA to G of DNA.

RNA nucleotides bond through the ________ of one to the __________of the adjacent nucleotide. The single ________ of RNA breaks away from the DNA and passes into the ___________ through a __________ in the nuclear membrane.

3 Use information in the text to construct a flowchart that shows translation. The first two stages have been completed for you.

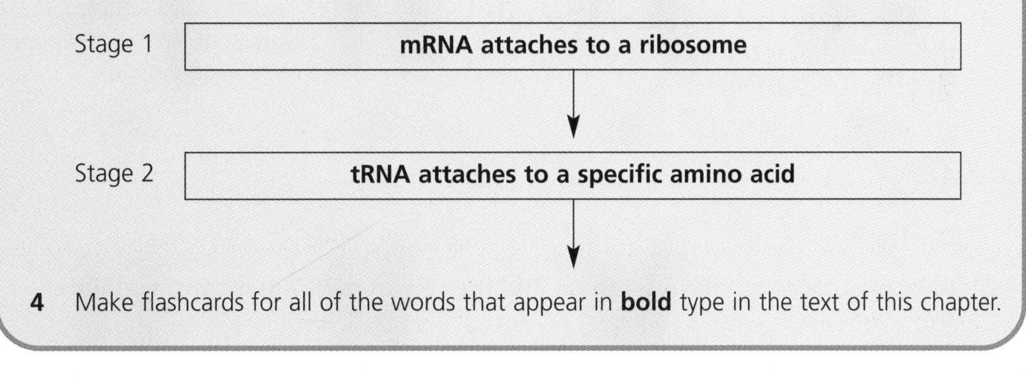

4 Make flashcards for all of the words that appear in **bold** type in the text of this chapter.

Exam Questions

1 Figure 3.16 shows a magnified view of part of a muscle.

The band marked X shows the presence of:

A myosin filaments on their own
B actin filaments on their own
C overlapping actin and myosin filaments
D the absence of actin and myosin filaments.

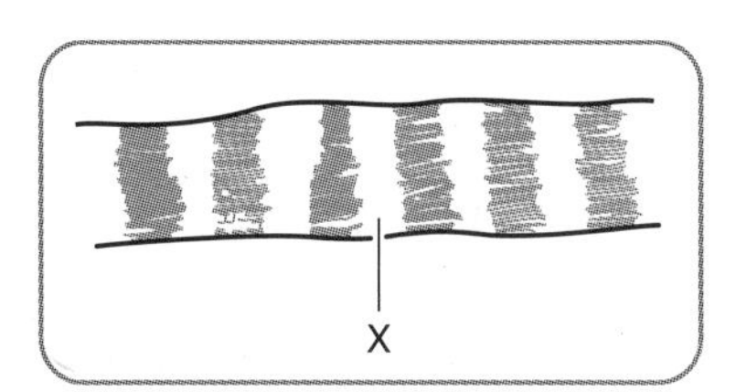

Figure 3.16 Part of a muscle

Exam Questions continued

2 Figure 3.17 represents part of a protein molecule.

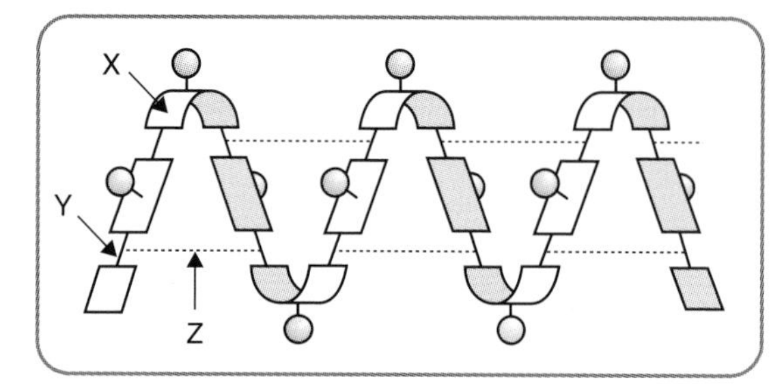

Figure 3.17 Part of a protein molecule

Molecule X is one of the molecules that form the chain.
Bonds Y and Z hold the protein together.
Name molecule X and identify bonds Y and Z. *(3)*

3 Figure 3.18 represents the arrangement of protein filaments in a relaxed muscle.

(a) Name filaments X and Y. *(1)*
(b) Describe what happens to these filaments when a muscle contracts. *(1)*
(c) Describe the change in the appearance of the bands after contraction. *(1)*

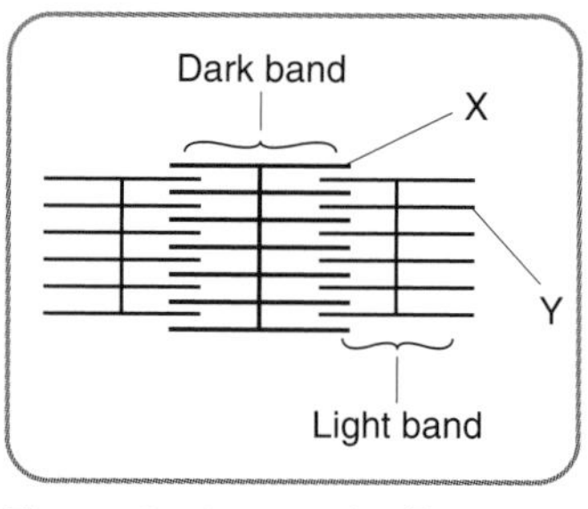

Figure 3.18 Protein filaments in a relaxed muscle

4 If 10% of the bases in a molecule of DNA are thymine, what is the ratio of adenine to cytosine bases in this DNA molecule?

A 1:1
B 1:2
C 1:3
D 1:4

5 Give an account of mRNA synthesis and the role of mRNA in protein synthesis. *(10)*

Answers

1 You should know that the light band contains filaments of actin only. **Answer** B

2 The diagram is of a protein and the basic units are **amino acids (X)**. **Peptide (Y)** bonds hold amino acids together in the chain and **hydrogen (Z)** bonds hold the protein shape together.

3 (a) You should know that X = myosin and Y = actin.
(b) This is the sliding filament theory: during contraction the filaments of actin and myosin slide in between one another.
(c) The dark bands stay the same width and the light bands become narrower.

Answers continued

4 If 10% are T then 10% are A (base pairing).
80% are C and G. Therefore 40% are C (base pairing)
Ratio of A to C = 10 : 40 Therefore = 1 : 4 **Answer** D

5

(8 marks for KU, 1 mark for coherence and 1 mark for relevance)

Nos	Mark scoring points		Comments
	An account of mRNA synthesis		***Write the sub-heading***
1	DNA unzips or hydrogen bonds broken	1	
2	Enzymes and ATP needed	1	
3	RNA nucleotides align with exposed bases of DNA	1	*Can show in a labelled diagram*
4	Base pairings **DNA** / **RNA**: A – U; T – A; G and C – C and G	1	
5	Nucleotides bond through sugar to phosphate	1	
6	mRNA transcribed from DNA	1	
			At least two points must be given
	The role of mRNA in protein synthesis		*Write the sub-heading*
7	mRNA passes out from the nucleus and attaches to a ribosome	1	*Can show in a labelled diagram*
8	tRNAs transport amino acids to ribosome	1	
9	Specific tRNA for each amino acid	1	
10	Triplets of bases on mRNA form codons *or* triplets of bases on tRNA form anticodons	1	*Can show in a labelled diagram*
11	Anticodons of tRNA match to codons on mRNA	1	
12	Amino acids bond by peptide bonds	1	
			At least two points must be given
	8 marks for knowledge and understanding		
	1 mark for coherence + 1 mark for relevance		***Maximum = 10 marks***

Answers continued

Coherence

1 **Sub-headings** and related information must be **grouped together**.
2 There must be a minimum of **five** correct points with at least two points from each. Both must apply correctly to gain the **coherence** mark.

Relevance

1 **Must not** give details of DNA replication.
2 **Must** give at least **two** relevant points from both mRNA synthesis and from the role in protein synthesis, and a minimum of **five** correct points. Both must apply correctly to gain the **relevance** mark.

Chapter 4

ENERGY TRANSFER

Key Ideas

Energy is stored in the bonds of glucose molecules as chemical energy. The energy stored in the chemical bonds of a glucose molecule is released within living cells in a series of enzyme-controlled reactions.

The breakdown of glucose to release energy is called respiration. In respiration, removal of hydrogen from a molecule is oxidation and addition of hydrogen is reduction.

The role and production of adenosine tri-phosphate (ATP) in energy transfer

The inter-conversion of adenosine di-phosphate (ADP), inorganic phosphate (Pi) and adenosine-tri-phosphate (**ATP**) is shown in Figure 4.1. Energy released from the breakdown of a glucose molecule is used to synthesise ATP. The ATP is synthesised when energy from the breakdown of glucose is used to bond Pi to a molecule of ADP.

The chemical energy is transferred from the bonds in glucose to the bonds in ATP. When ATP breaks down to re-form ADP and Pi, energy is released. The energy released from the breakdown of ATP is the immediate source of energy for all cellular activities.

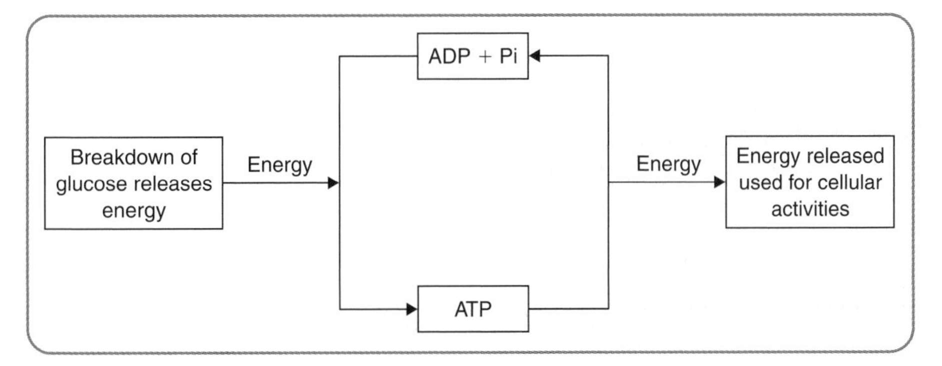

Figure 4.1 Interconversion of ADP, Pi and ATP

The mass of ATP present in the body of an adult human is about 50 g. ATP is produced only when it is needed and the rate of ATP synthesis is equal to the rate at which it is used up.

Hints and Tips

Some of the energy released during respiration is lost as heat energy.

The immediate source of energy for all cell activities is ATP.

Cellular activities include muscle contraction, cell division, synthesis of protein and transmission of nerve impulses.

In experimental situations, if the temperature increases it is due to the release of heat energy in respiration.

For Practice

1 Describe the synthesis of ATP. glucose —energy→ ATP
2 Name the immediate source of energy for all cell activities. breakdown of ATP

Aerobic respiration

Respiration occurs in stages controlled by enzymes. The first stage of respiration is **glycolysis.** In glycolysis 6-carbon glucose (6C) is broken down to two molecules of pyruvic acid (3C). Study the pathway of glycolysis as outlined in Figure 4.2. Energy from ATP is required to initiate the process. ATP is then synthesised from the energy released during breakdown to give a net gain of **2 ATPs**. Glycolysis takes place in the **cytoplasm** of the cell.

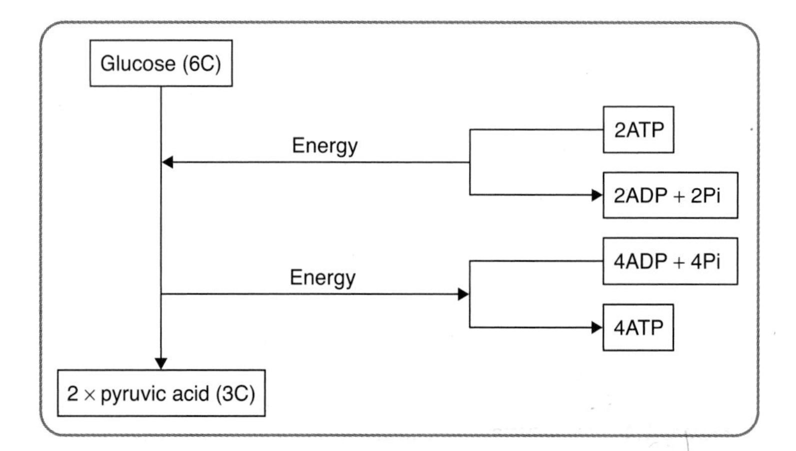

Figure 4.2 Glycolysis

Further breakdown of pyruvic acid occurs in organelles named the mitochondria (Figure 4.3). The mitochondrion has a double membrane. The inner membrane is greatly folded to give a **large surface area**. The presence of a large surface area increases the activity of a mitochondrion.

The folds are called **cristae** (site of cytochrome system). The central cavity is filled with fluid and contains enzymes. The fluid is called the **matrix** (site of Krebs cycle).

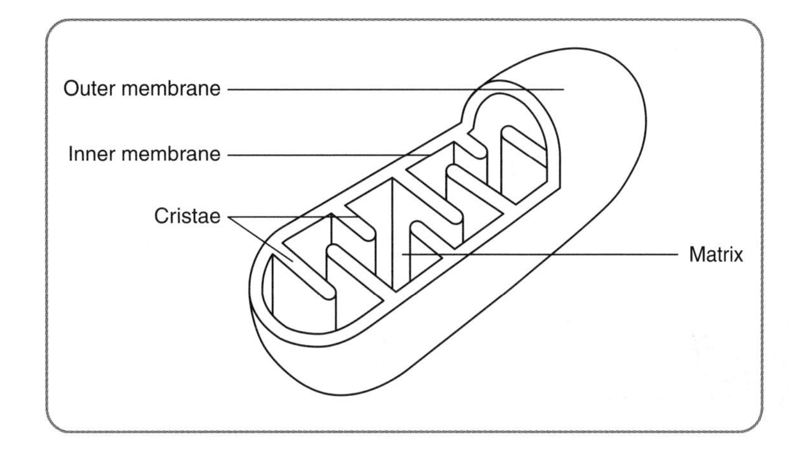

Figure 4.3 Mitochondrial structure

The second stage of aerobic respiration is Krebs cycle (Figure 4.4). A molecule of 3C-pyruvic acid enters a mitochondrion. The 3C-pyruvic acid is broken down to form a 2C-acetyl group and a carbon is lost as carbon dioxide. The 2C-acetyl group bonds with co-enzyme A (CoA) to form acetyl-CoA. A 2C-acetyl group bonds to a 4C-compound to form a 6C-compound called **citric acid**. The CoA molecule is free to collect and transport another acetyl group to Krebs cycle. 6C-citric acid is broken down in stages to reform the original 4C-compound. Each time that a carbon is lost it is lost as carbon dioxide. **Hydrogen** is released during Krebs cycle. The hydrogen released is picked up by a co-enzyme named NAD and **reduced co-enzyme** ($NADH_2$) is formed.

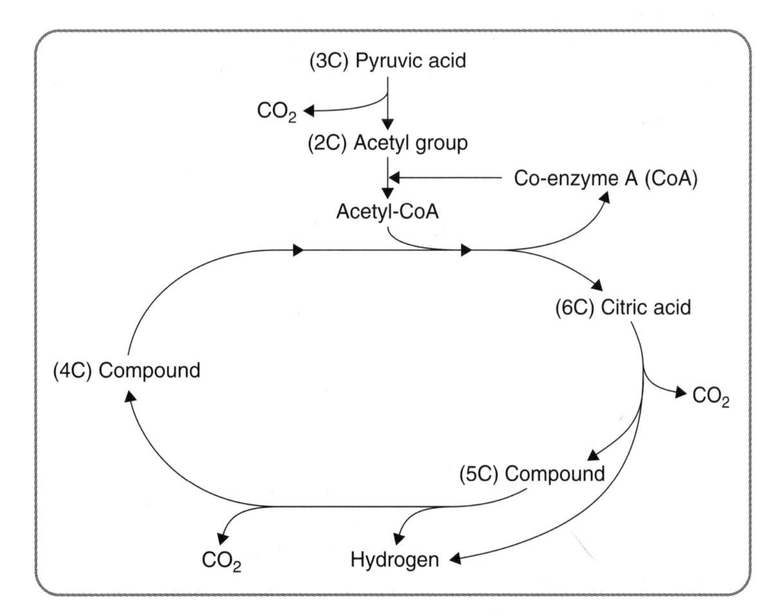

Figure 4.4 Krebs cycle

The third stage of aerobic respiration is the cytochrome system (Figure 4.5).

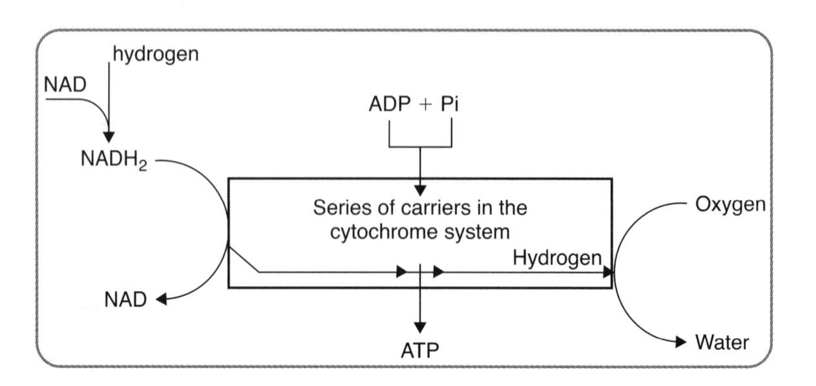

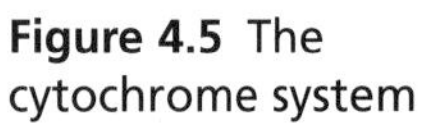
Figure 4.5 The cytochrome system

Hydrogen is transported from Krebs cycle to the cytochrome system by $NADH_2$. The cytochrome system consists of a series of carriers. Hydrogen and electrons are transferred through the **carrier system**. Energy is released during the transfer of hydrogen and electrons. The energy released is used to synthesise ATP from ADP and inorganic phosphate. The final hydrogen acceptor is **oxygen**. Hydrogen and oxygen bond to form water.

In aerobic respiration, a total of 38 ATPs are synthesised per glucose molecule, 2 ATPs are synthesised from glycolysis and 36 ATPs from the cytochrome system.

The equation for aerobic respiration is:

Glucose + oxygen + 38ADP + 38Pi ⟶ Carbon dioxide + water + 38ATP

Hints and Tips

Enzymes control respiration, so changes in temperature have the same effects on the rate of respiration as on enzyme action.

Remember that respiration is *not* breathing. Breathing gets the oxygen into the body and the oxygen is needed for aerobic respiration.

Lysis means 'to split' and *glyco* refers to glucose. Glycolysis is the splitting of glucose.

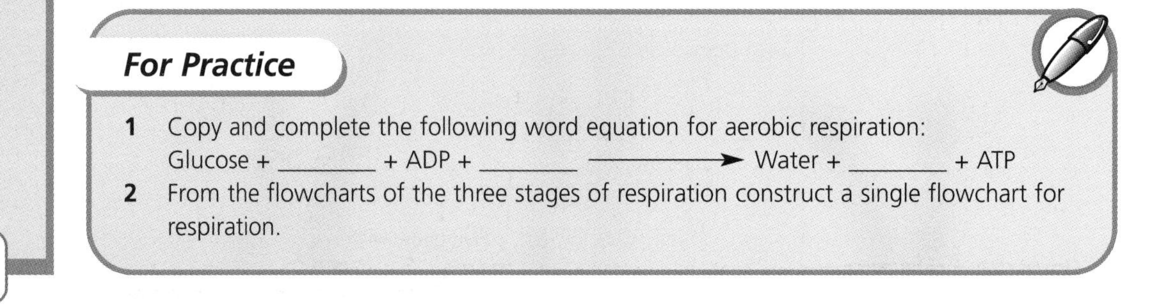

For Practice

1. Copy and complete the following word equation for aerobic respiration:
   Glucose + ________ + ADP + ________ ⟶ Water + ________ + ATP
2. From the flowcharts of the three stages of respiration construct a single flowchart for respiration.

Aerobic and anaerobic pathways of respiration

Aerobic respiration takes place in the presence of oxygen.

Anaerobic respiration takes place in the absence of oxygen. Figure 4.6 shows the pathway of anaerobic respiration in human muscle cells.

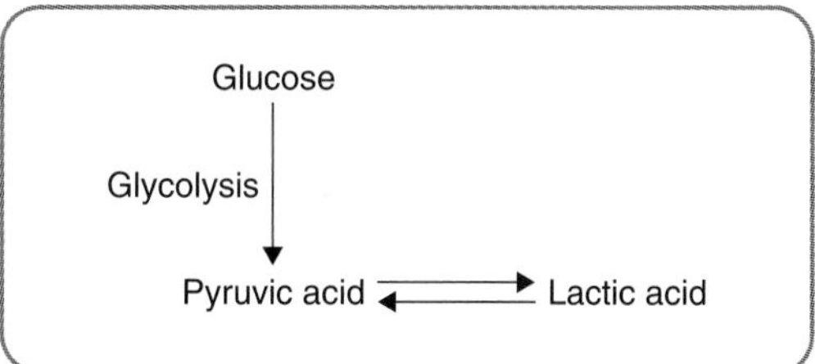

Figure 4.6 Anaerobic respiration

When oxygen is not available in the tissues this is called an oxygen debt and conditions are anaerobic. In oxygen debt pyruvic acid is converted to **lactic acid**. When oxygen becomes available in the tissues the oxygen debt is paid back and lactic acid is converted back to pyruvic acid. Aerobic respiration yields 38 molecules of ATP per glucose molecule and anaerobic respiration yields 2 molecules of ATP per glucose molecule. Aerobic respiration is 19 times more efficient than anaerobic respiration. Anaerobic respiration takes place in the **cytoplasm.**

Hints and Tips

*A**n**aerobic* has a letter ***N***. N = ***NO*** = **N**o **O**xygen present.

Glycolysis is a stage common to both aerobic and anaerobic respiration.

In human tissues conversion of pyruvic acid to lactic acid is reversible.

An increase in lactic acid leads to muscle fatigue (muscles are less efficient).

You must know the numbers of carbon atoms in:
glucose, citric acid, pyruvic acid, lactic acid and the acetyl group.

For Practice

1 Copy and complete the word equation for anaerobic respiration in human muscles:
Glucose ⟶ ____________________ + ATP

2 Build in the pathway for anaerobic respiration on your respiration flowchart.

***For Practice** continued* ➢

For Practice *continued*

3 What is the effect of an oxygen debt in muscle tissue?
4 Why is aerobic respiration described as being 19 times more efficient than anaerobic respiration?
5 Copy and complete the table below to compare aerobic and anaerobic respiration.

Type of respiration	Location in cell	Number of ATPs produced	Final metabolic product(s) other than ATP
Aerobic			
Anaerobic			

Sources of energy within the cell

Carbohydrates, lipids (fats) and proteins are sources of energy within living cells. Figure 4.7 represents carbohydrate structure.

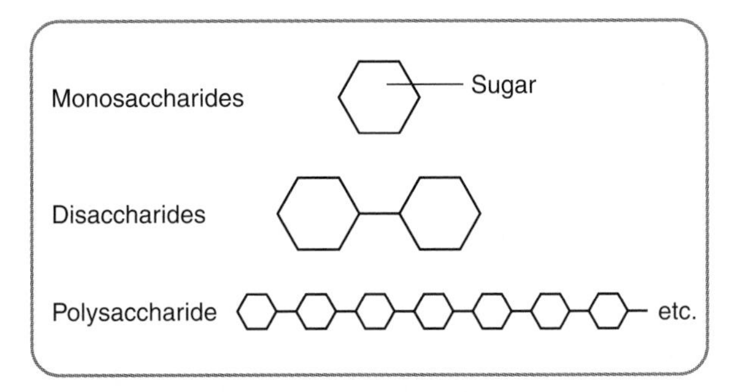

Figure 4.7 Carbohydrate structure

The term *saccharide* refers to sugar. Monosaccharides are molecules made up of one sugar unit. Examples include glucose, deoxyribose and ribose. Sugar molecules built up from two sugar units are disaccharides. Examples include sucrose, maltose and lactose. Molecules built up from many sugar units are polysaccharides. Examples include glycogen and starch. Polysaccharides such as glycogen and starch are energy stores within the organism. In humans, glycogen is stored in the liver and muscle tissues. Different energy sources are utilised under different conditions.

1 During **prolonged exercise**, as in a marathon, glucose is constantly absorbed from the blood by muscles and used in respiration. Glucose is replaced by the release of glucose into the blood from the breakdown of stored glycogen.
2 During prolonged exercise, as in a marathon, fat starts to be used as an energy source as the stores of carbohydrate become used up.
3 During **prolonged starvation** protein is used as an energy source after the stores of glycogen and fats are used up.

Role of lipids within the body

Lipids act as a source of stored energy within the body. Lipids stored beneath the skin act as an insulation layer that prevents heat loss.

The myelin sheath of a neurone consists of fat-containing cells that insulate the axon from electrical activity. The insulation acts to increase the rate of transmission of impulses along the axon (see Chapter 13 The nervous system).

During movement fat pads in the ball and heel of the feet act as shock absorbers and give physical protection to the underlying bones. Fat pads in the hand also give physical protection to the underlying bones.

Lipids are used in the transport of fat-soluble vitamins and hormones.

Some hormones are lipids. Examples include testosterone, oestrogen and progesterone.

For Practice

1. Name three possible sources of energy within living cells.
2. Describe a polysaccharide.
3. Which energy sources are used during prolonged exercise?
4. Which energy source is used during prolonged starvation?
5. Give two examples of lipids being used for insulation.
6. Copy and complete these sentences on the role of lipids in the body.
   Lipids are used in the ___________ of hormones and ___________. When landing on your feet __________ ___________ act as ________ absorbers.
7. Make flashcards for all of the words that appear in **bold** type in the text of this chapter.

Exam Questions

1 Liver tissue of mass 4 g uses up 3.0 cm^3 of oxygen in 5 minutes. Calculate the rate of respiration of liver tissue in terms of cm^3 of oxygen used per gram of liver tissue per minute.

A 0.010
B 0.015
C 0.100
D 0.150

2 This list shows the role of chemicals within cells:

1 Nerve insulation
2 Storage of energy
3 Transport of hormones
4 Transport of oxygen

Exam Questions continued

Which of the following are roles of lipids?

A 1, 2 and 3 only
B 1, 3 and 4 only
C 2, 3 and 4 only
D 1, 2, 3 and 4

3 Figure 4.8 represents stages of aerobic respiration in a liver cell.

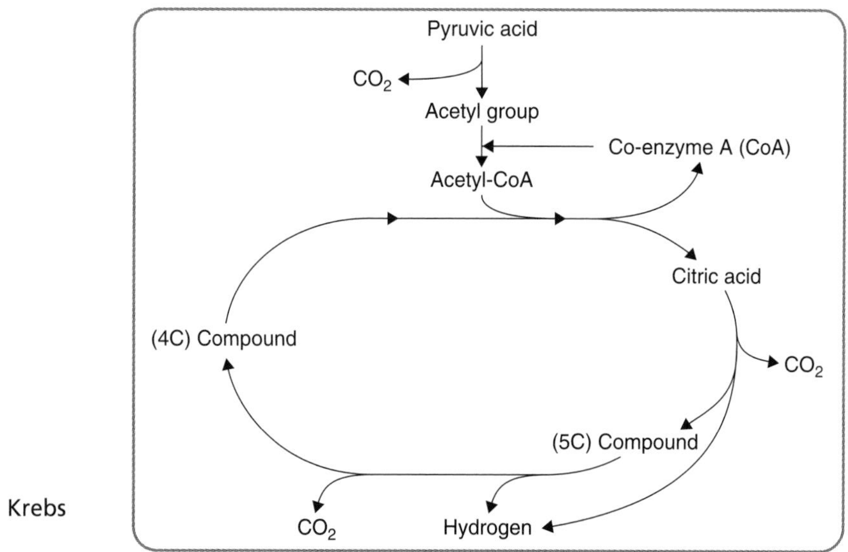

Figure 4.8 Krebs cycle

(a) Use molecules named in the diagram to complete the table. *(2)*

Number of carbon atoms	Molecule(s)
6	
3	
2	

(b) State the exact location within a liver cell of:
(i) Glycolysis
(ii) Krebs cycle
(iii) the cytochrome system. *(2)*

(c) Apart from glucose, name two other substances that must be present for glycolysis to take place. *(1)*

(d) Name the molecule formed during the passage of hydrogen through the series of hydrogen carriers. *(1)*

(e) Why does the cytochrome system stop when oxygen is absent? *(1)*

(f) Describe what happens to pyruvic acid in liver cells in the absence of oxygen. *(1)*

Answers

1 Answer to be in cm^3/g/minute.
From figures given this is expressed as 3.0 cm^3/4 g/5 minutes.
To find volume for 1 g divide 3.0 by 4 = 0.75 cm^3/g/5 minutes.
To find volume for 1 minute divide 0.75 by 5 = 0.150 cm^3/g/minute.
Answer D

2 You should know that haemoglobin transports oxygen and that the other three are roles of lipids. **Answer** A

3 (a)

Number of carbon atoms	Molecule(s)
6	Citric acid
3	Pyruvic acid
2	Acetyl group

You must know the number of carbon atoms in each molecule in respiration.
3 correct = 2 marks; 2 correct = 1 mark

(b) (i) Glycolysis is located in cytoplasm.
(ii) Krebs cycle is in the matrix of mitochondrion.
(iii) the cytochrome system is in the cristae of mitochondrion.
3 correct for 2 marks; 2 correct for 1 mark
You are penalised once if you do not mention the mitochondrion.

(c) You should know that ATP is required to initiate glycolysis and that the breakdown is in steps and each step requires an enzyme.
Answer ATP and enzymes

(d) You should know that ATP is synthesised as hydrogen passes through carriers.

(e) You should know that oxygen is needed to remove the hydrogen.
Answer When oxygen is absent hydrogen is not removed and the build-up of hydrogen stops the cytochrome system.

(f) You should know that in the absence of oxygen conditions become anaerobic and that pyruvic acid is converted to lactic acid in anaerobic conditions.
Answer Pyruvic acid is converted to lactic acid.

Chapter 5

CELL TRANSPORT

Key Ideas

Membranes form part of the structure of cells and cell organelles. Membranes have several functions and these include control of transport of substances into and out of cells and cell organelles. Substances are transported by diffusion, osmosis and active transport.

Structure and function of membranes

Identify membrane structures by referring to Figure 5.1.

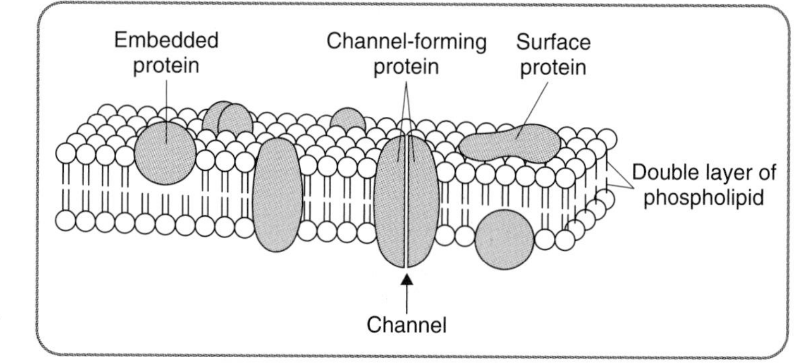

Figure 5.1 Membrane structures

The membrane has a double layer of phospholipid. **Proteins** are embedded in the **phospholipid** bilayer. Some proteins form as surface proteins; some proteins are embedded within the membrane; and some proteins form channels that extend across the phospholipid bilayer. Phospholipids and proteins are in constant motion. This is described as a **fluid mosaic model**. The phospholipid bilayer allows transport of fat-soluble molecules across the membrane. These include vitamins and hormones. The **channels** allow transport of water-soluble molecules across the membrane, so the membrane is described as being **porous**. Some surface proteins act as antigens that identify the cell as 'self' and some act as receptors for hormones. The rough endoplasmic reticulum (ER) and the Golgi apparatus are structures formed from membranes. These structures are involved in the transport and secretion of proteins.

The transport and secretion of substances within a cell by rough ER and Golgi apparatus are outlined in Figure 5.2.

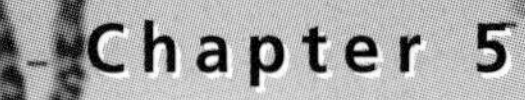

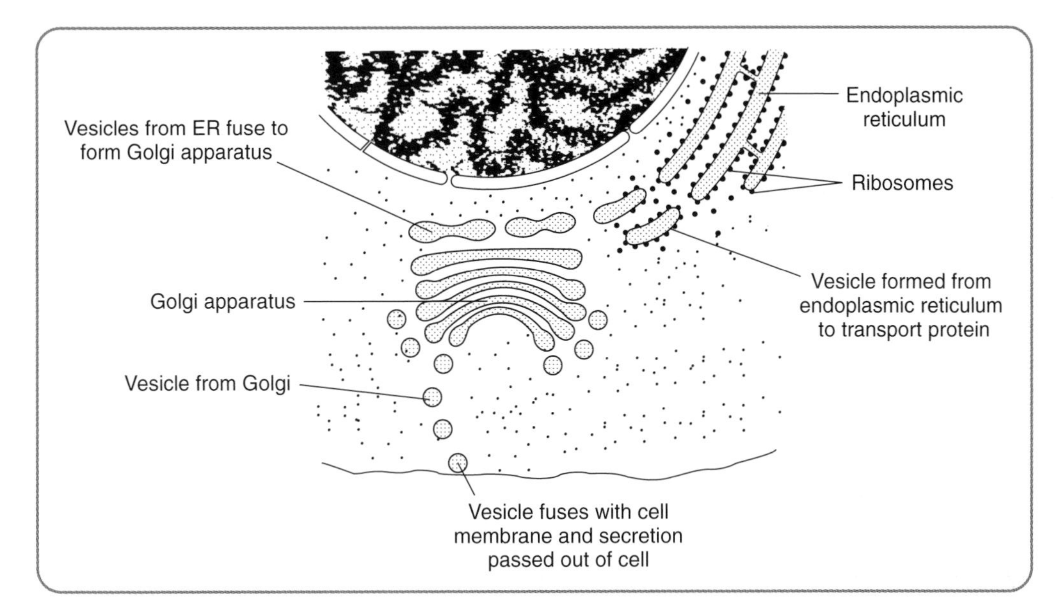

Figure 5.2 Transport and secretion of substances within a cell

The **rough endoplasmic reticulum** (ER) forms as tube-like structures with ribosomes attached. Proteins synthesised on the ribosomes pass into the cavity of the rough ER. Membranes of the rough ER pinch off to form a vesicle (small vacuole) which transports the protein to the **Golgi apparatus**. The protein is processed within the Golgi apparatus and made ready for secretion. Membranes of the Golgi apparatus pinch off to form **vesicles** which travel to the cell membrane. Vesicles fuse with the cell membrane and the contents are secreted out of the cell. Substances secreted from cells are hormones or enzymes. Examples include secretion of insulin by the pancreas and secretion of enzymes into the digestive tract. Some vesicles containing digestive enzymes remain inside the cell and develop into organelles named **lysosomes**.

Smooth ER is similar in structure to rough ER except there are no ribosomes on the surface. Smooth ER varies in function within different tissues. Smooth ER is involved in the storage of enzymes, in the metabolism of glycogen in liver cells and in the metabolism of hormones in cells of the gonads.

Hints and Tips

Never refer to the membrane as lipid; always as *phospholipid*.

Ribosomes are present on the surface of rough ER but not on that of smooth ER.

Large molecules cannot pass through the membrane. Soluble substances enter through channels in proteins and fat-soluble substances pass through the phospholipid bilayer.

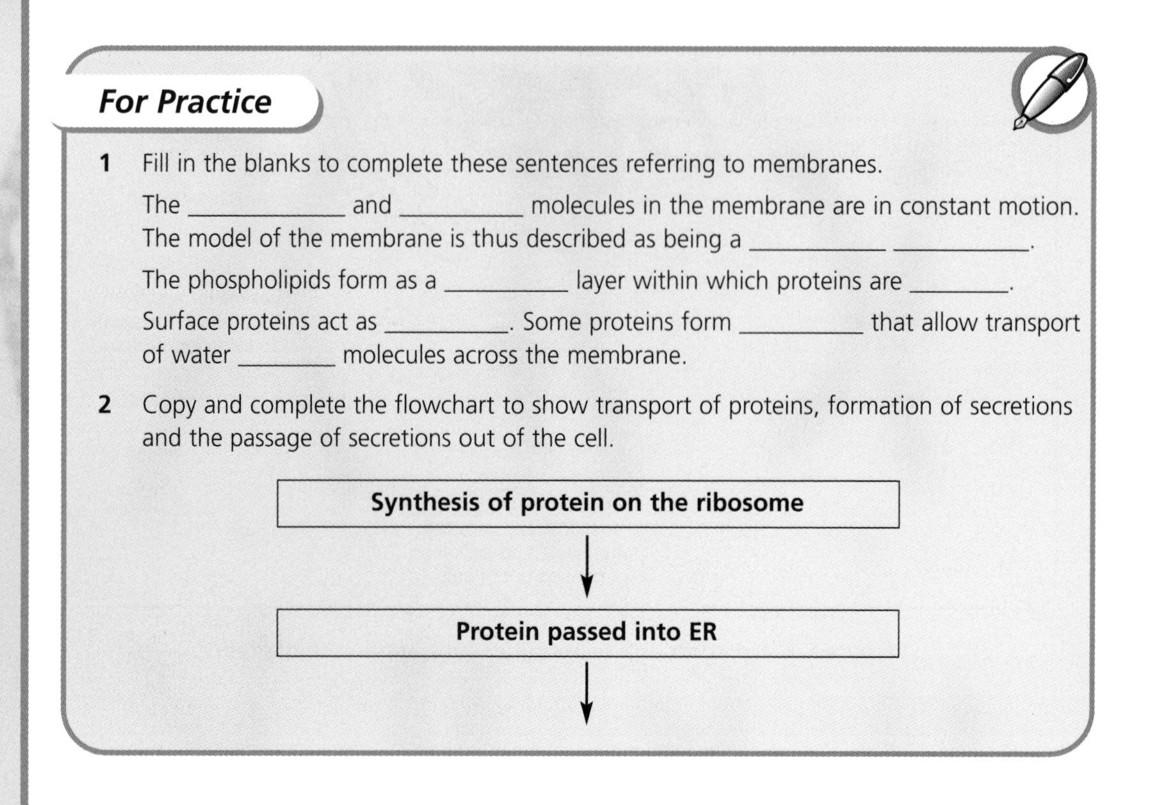

For Practice

1 Fill in the blanks to complete these sentences referring to membranes.

The ____________ and __________ molecules in the membrane are in constant motion. The model of the membrane is thus described as being a ___________ ___________.

The phospholipids form as a __________ layer within which proteins are ________.

Surface proteins act as __________. Some proteins form __________ that allow transport of water ________ molecules across the membrane.

2 Copy and complete the flowchart to show transport of proteins, formation of secretions and the passage of secretions out of the cell.

Synthesis of protein on the ribosome

↓

Protein passed into ER

↓

Diffusion, osmosis and active uptake

The cell membrane controls movement of some materials into and out of the cell. Such a membrane is described as being a **selectively permeable membrane** (SPM).

In a solution the solute is the substance dissolved, and water is the solvent. Diffusion of molecules occurs if there is a solute concentration difference between two areas.

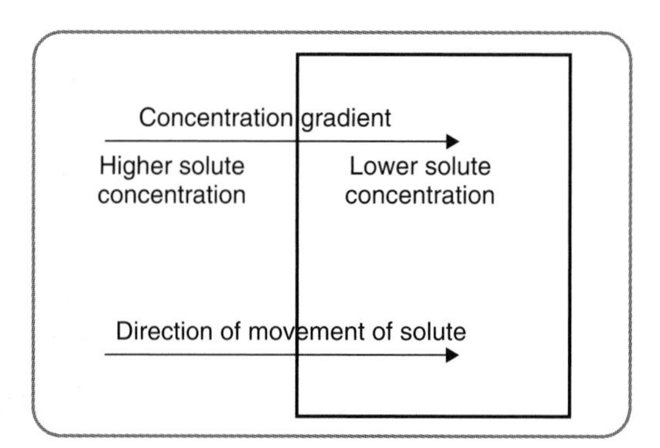

Figure 5.3 Movement of solute

Figure 5.3 shows that the area outside the cell has a higher solute concentration than the area inside. This solute concentration difference is described as a solute concentration gradient and is shown by an arrow. The net movement of solute down the solute concentration gradient is the process named diffusion.

Diffusion is the movement of molecules from an area of higher solute concentration to an area of lower solute concentration down a solute concentration gradient. Osmosis is a 'special case' of diffusion. Osmosis refers to the diffusion of water molecules. The rules for diffusion of water are the same as for diffusion of solutes. There has to be a difference in water concentration between two areas. The areas are separated by a SPM (cell membrane) and one area has a higher water concentration than the other.

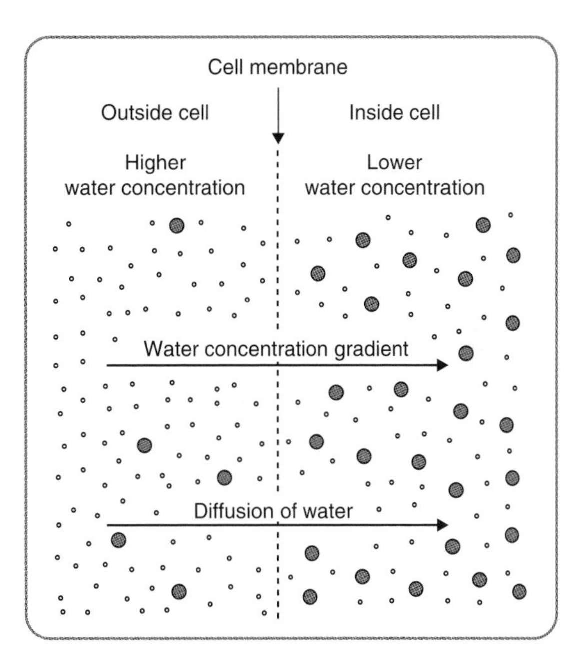

Figure 5.4 Water concentration gradient

Figure 5.4 shows that the area outside the cell has a higher water concentration than the area inside the cell. This water concentration difference is described as a water concentration gradient and is shown by an arrow. The net movement of water down the water concentration gradient is the process named osmosis. **Osmosis** is the overall diffusion of water from an area of higher water concentration to an area of lower water concentration across a SPM.

Osmosis has effects on the cells of the body. Figure 5.5 identifies the effects of different water concentrations on red blood cells (RBCs). In pure water (higher water concentration than cell contents) water enters the cell by osmosis. The cell swells and bursts. In a 1.7% salt solution (lower water concentration than cell contents) the cell loses water by osmosis and shrinks. In solution with a water concentration the same as the cell contents there is no overall osmotic effect and the cell remains as normal.

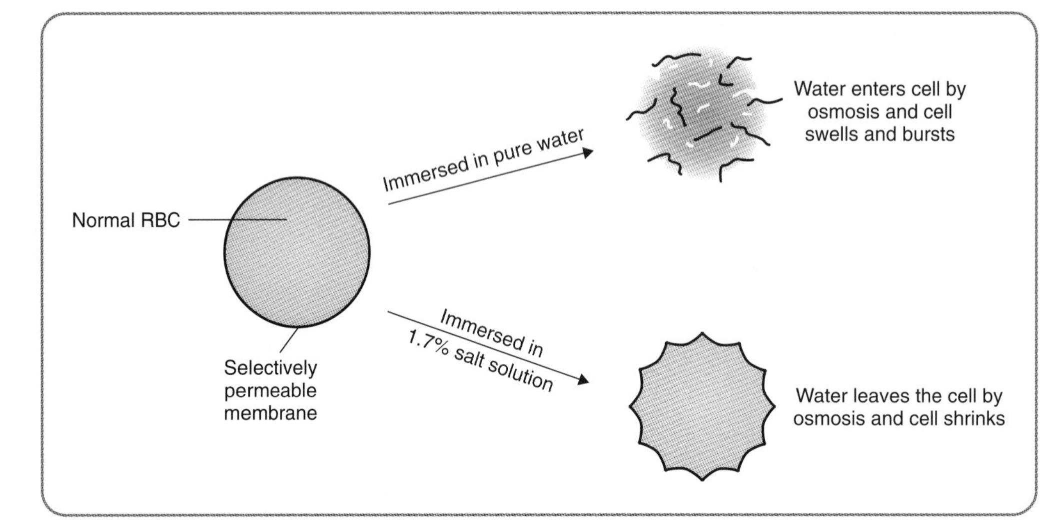

Figure 5.5 Effects of osmosis on RBCs

Hints and Tips

The lower the solute concentration the higher is the water concentration.

A 1% solution has a higher water concentration than a 5% solution.

Always make it clear which concentration you are dealing with. Write *water concentration* if it is osmosis and *solute concentration* if it is diffusion.

For Practice

1 Copy and complete the table below to show the pathway of diffusion of glucose, and amino acids, and the importance of this to the cell.

Substance diffusing	Pathway	Importance to the cell
Glucose	Into cell	
Amino acids		

2 State the direction of osmosis in the following situations.

Situation	Concentration outside the cell	Concentration inside the cell
A	1.1% solution	2.2% solution
B	2% solution	2% solution

Chemical ions can be taken into and out of cells against the concentration gradient. This is **active transport**. Active transport is selective and requires energy. ATP is the immediate source of the energy required.

The process of active transport is shown in Figure 5.6. Proteins embedded in the phospholipid bilayer act as carriers. A carrier protein picks up an ion from the extracellular fluid. The ion is present in a concentration lower than that of the intracellular fluid. ATP is the immediate source of energy that allows the carrier protein to rotate. The ion is released into the intracellular fluid.

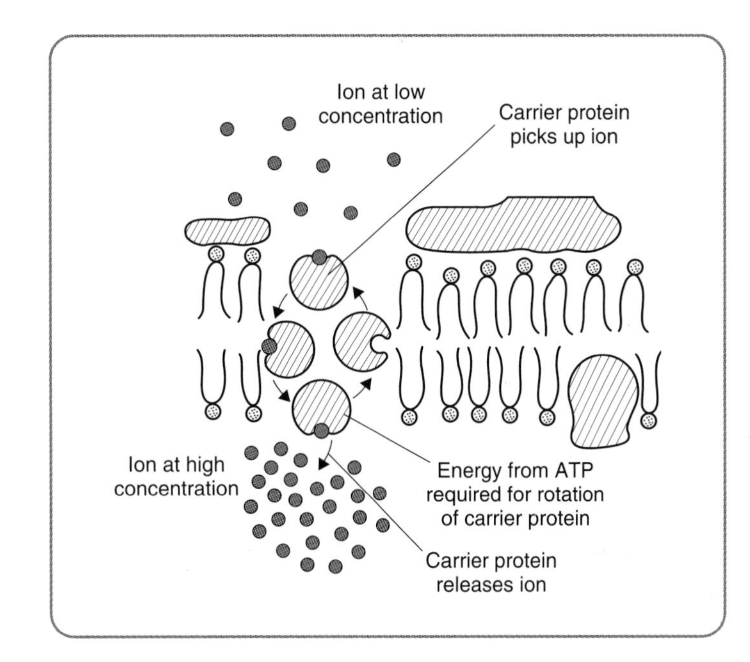

Figure 5.6 Active transport

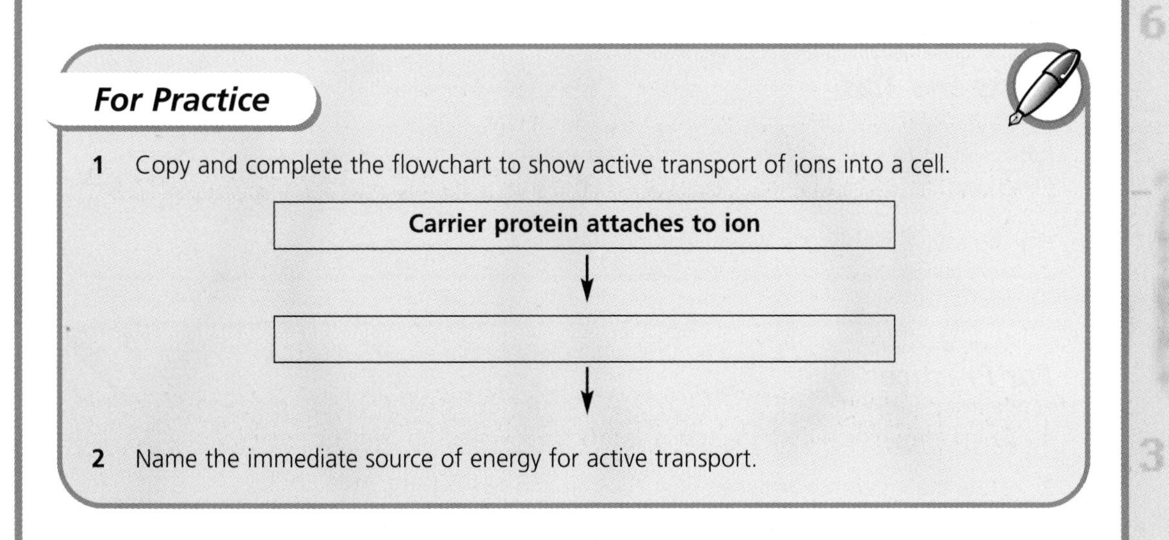

For Practice

1 Copy and complete the flowchart to show active transport of ions into a cell.

Carrier protein attaches to ion
↓
↓

2 Name the immediate source of energy for active transport.

Endocytosis and exocytosis

Substances can enter into or pass out of the cell by large movements of the cell membrane. **Endocytosis** is the process by which a cell engulfs materials and takes them into the cell from the surrounding fluid.

There are two types of endocytosis: **phagocytosis** and **pinocytosis.** The difference is in the size of the materials taken into the cell.

In phagocytosis it is relatively large structures such as bacteria that are taken into the cell. In pinocytosis it is mainly fluid. Figure 5.7 outlines the stages of phagocytosis. The cell membrane starts to surround the bacteria and the bacteria are enclosed in a vacuole. Lysosomes attach to the vacuole and enzymes from the lysosomes digest the bacteria. In pinocytosis the cell membrane surrounds the extracellular fluid and encloses it in a vacuole.

Exocytosis is the process by which a cell passes material out of the cell into the surrounding fluid. Refer to the function of the Golgi apparatus (Chapter 5, page 39). Hormones or enzymes are enclosed within a vesicle. The vesicle is moved to the surface of the cell and fuses with the membrane. The hormone or enzyme is secreted out of the cell.

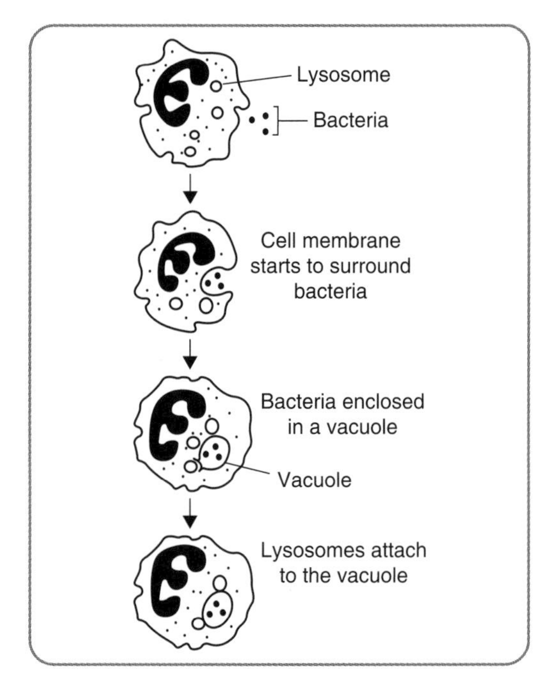

Figure 5.7 Phagocytosis

Hints and Tips

Exo is similar to 'exit'. In exocytosis materials exit the cell. If you remember one term you remember the other. In endocytosis they do not exit; materials enter.

Vesicles can be considered to be small vacuoles.

For Practice

1 Using information from the text, construct a flowchart to describe phagocytosis.

2 State the difference between phagocytosis and pinocytosis.

3 Make flashcards for all of the words that appear in **bold** type in the text of this chapter.

Exam Questions

1 Insulin is a protein synthesised in cells of the pancreas and secreted into the blood. The following are stages in this process:

1 protein transported in endoplasmic reticulum
2 protein modified in Golgi apparatus
3 mRNA attaches to a ribosome
4 protein transported from Golgi apparatus
5 mRNA transcribed.

The correct sequence of these stages is:

A 3 5 2 4 1
B 3 5 1 2 4
C 5 3 2 4 1
D 5 3 1 2 4.

2 Which of the following are examples of endocytosis?

A Phagocytosis and diffusion
B Phagocytosis and pinocytosis
C Diffusion and osmosis
D Phagocytosis, diffusion, osmosis and pinocytosis.

3 Figure 5.8 represents the structure of the plasma membrane. The membrane is described as having a porous and fluid nature.

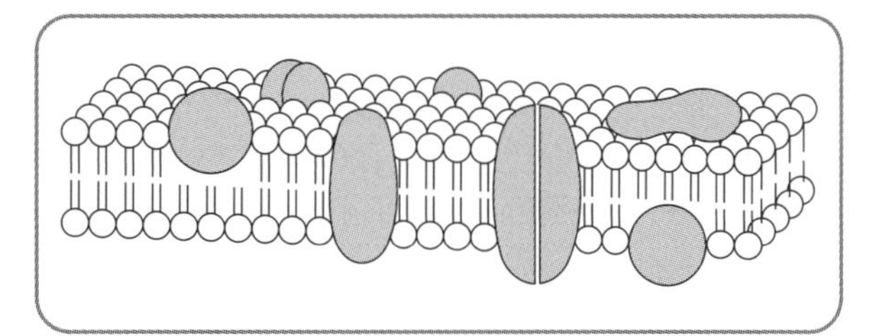

Figure 5.8 Plasma membrane

(a) Label with the letter P the structure that is composed of phospholipids. *(1)*
(b) The average speed of movement of a phospholipid is 2 μm per second. How far would a phospholipid travel in 3 minutes? *(1)*
(c) Explain why membranes are described as having a porous nature. *(1)*
(d) The concentration of sodium ions in the fluid outside the cell is 150 mmol/l. Inside the cell the concentration of these ions is approximately 25 mmol/l. Explain how this difference in concentration is maintained. *(2)*

Exam Questions continued

4 Figure 5.9 represents structures present in a white blood cell.

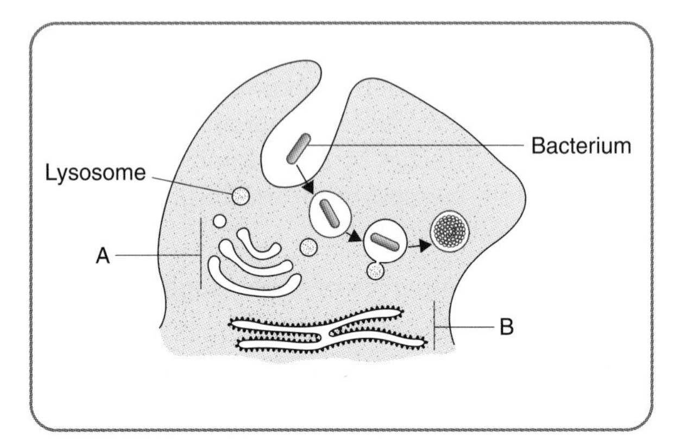

Figure 5.9 White blood cell

(a) The cell has engulfed a bacterium. What name is given to this process? *(1)*
(b) Describe the role of lysosomes in this process. *(2)*
(c) Name structures A and B and describe their roles in the cell. *(2)*

Answers

1 This question integrates KU from protein synthesis and KU from secretion. You should know that the process starts at transcription, mRNA attaches to a ribosome, then transport to Golgi, modification and then secretion. **Answer** D

2 *Endo*cytosis – not *exo*cytosis – therefore taking materials in. Examples are phagocytosis and pinocytosis. **Answer** B

3 (a) Use a vertical bracket so as to include both layers. Place letter P as instructed.
(b) 3 minutes = 180 seconds. **Answer** $3 \times 180 =$ **360** μm (include units).
(c) Because there are channels across the membrane and molecules can pass through them.
(d) Because sodium ions are pumped out (1st mark) and this requires energy for movement against the concentration gradient (2nd mark).

4 (a) You should know that engulfing of bacteria is phagocytosis.
(b Lysosomes attach and fuse with the vacuole (1st mark) and the enzymes of lysomes digest the bacterium (2nd mark).
(c) A – Golgi apparatus used to process materials ready for secretion.
B – Rough endoplasmic reticulum used to transport protein.

Chapter 6

CELLULAR RESPONSE IN DEFENCE

Antigens are protein molecules present on the surface membrane of all cells. Figure 6.1 shows antigens present on the surface of the influenza virus. The cellular defence system of our body, the immune system, responds to antigens and destroys the antigen-carrying cells. However, a central feature of our immune system is the ability to distinguish between 'self' and 'foreign' antigens. Every individual, other than identical twins, has a unique antigen 'signature' that is recognised by our immune system as being 'self' and thus our cells are not destroyed. Breakdown of the recognition system leads to **autoimmune** disorders such as rheumatoid arthritis, multiple sclerosis and insulin-dependent diabetes mellitus.

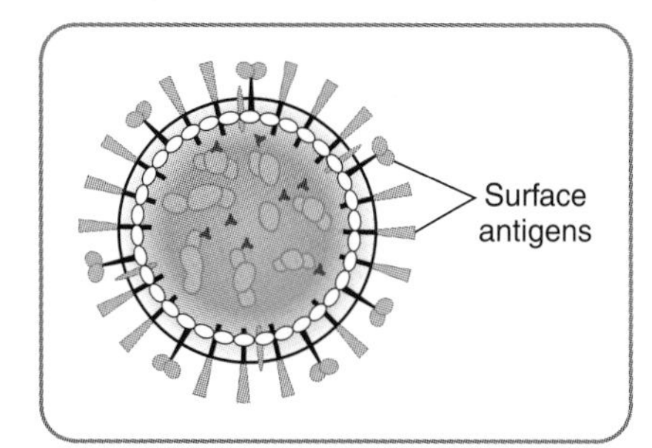

Figure 6.1 Antigens on influenza virus

Natural immunity

Natural immunity is the ability of the body to recognise foreign material and mobilise white blood cells (WBCs) and cell products to deal with the material.

Figure 6.2 shows the development of three different types of WBC involved in the immune system.

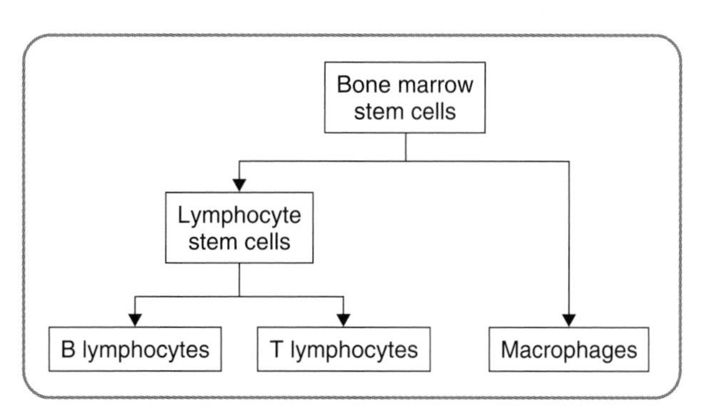

Figure 6.2 Development of three types of white blood cell

All cells develop from stem cells in the bone marrow.

Macrophages are phagocytes. Their function is to surround bacteria or viruses and enclose them in a vacuole. Lysosomes attach to the vacuole and enzymes from the lysosomes digest the bacteria (see Chapter 5 page 44).

B lymphocytes produce Y-shaped protein molecules in response to the surface antigens of bacteria and viruses. These proteins are antibodies. Production of cell products such as antibodies in response to an antigen is called the **humoral response**. The stages in the humoral response shown by B lymphoctes are outlined in Figure 6.3.

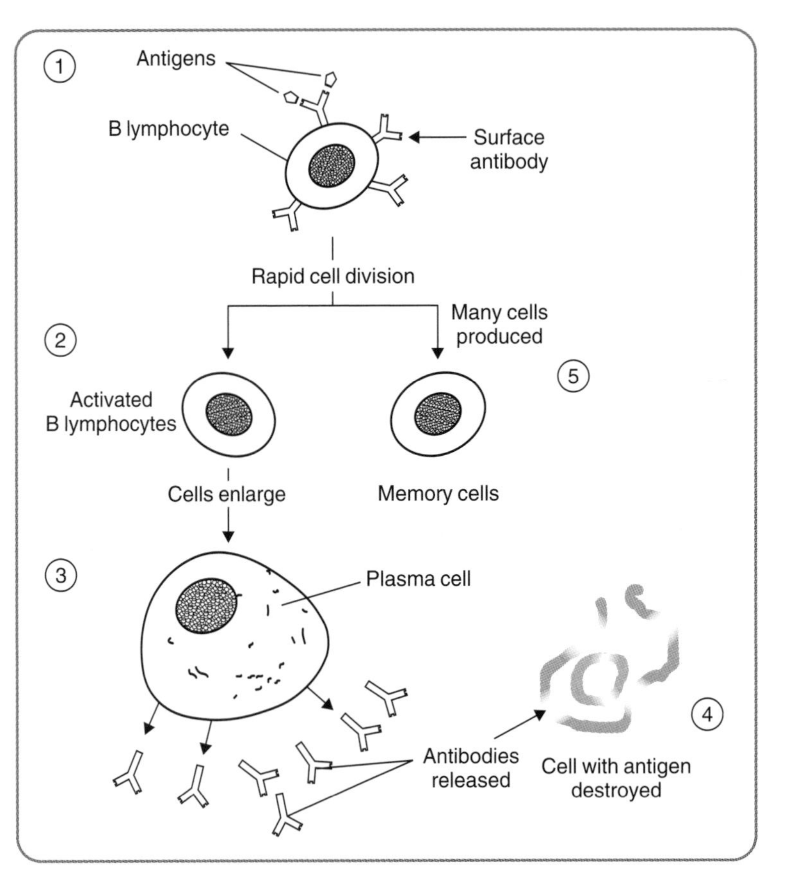

Figure 6.3 Stages in humoral system

Stage 1 The B lymphocyte comes into contact with a specific antigen.

Stage 2 The B cell divides rapidly to form a group of identical cells.

Stage 3 Some cells develop into plasma cells that produce the antibody specific to the antigen.

Stage 4 The released antibody leads to the destruction of cells with the antigen.

Stage 5 Other cells develop into memory cells.

If the same antigen enters the body again, the **memory cells** quickly produce more of the plasma cells.

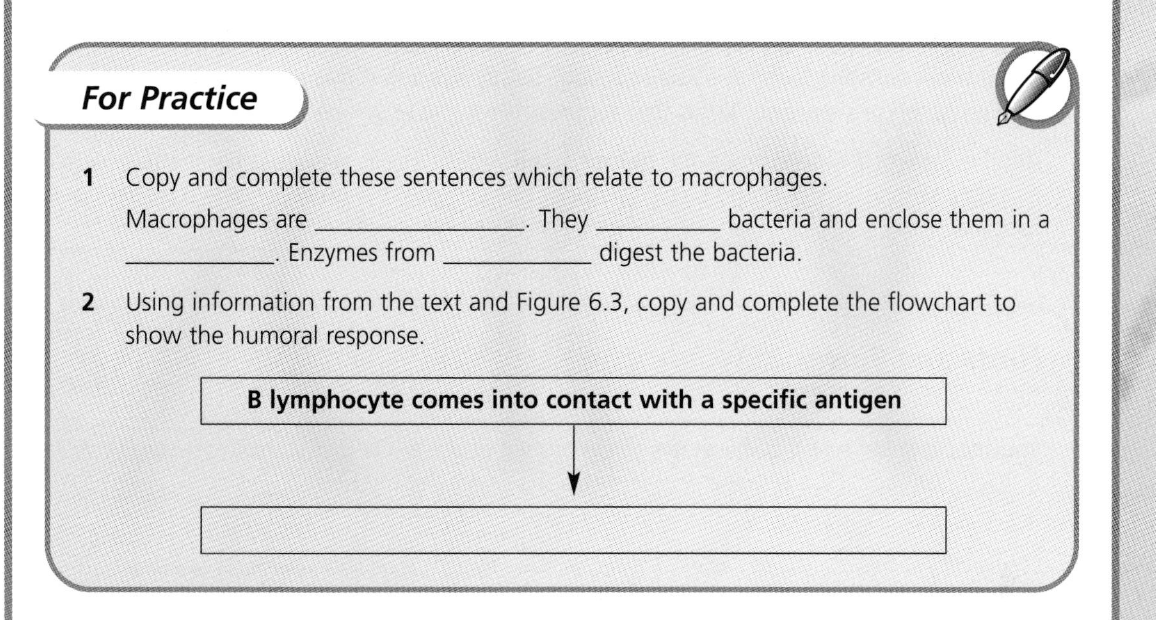

For Practice

1 Copy and complete these sentences which relate to macrophages.

Macrophages are ________________. They __________ bacteria and enclose them in a ____________. Enzymes from ____________ digest the bacteria.

2 Using information from the text and Figure 6.3, copy and complete the flowchart to show the humoral response.

B lymphocyte comes into contact with a specific antigen

↓

[ ]

A type of T lymphocyte recognises cells infected by a virus or bacteria and kills them. These are **killer T lymphocytes**. The destruction of infected cells by killer T lymphocytes is called the **cell-mediated response**. The stages in the cell-mediated response are outlined in Figure 6.4.

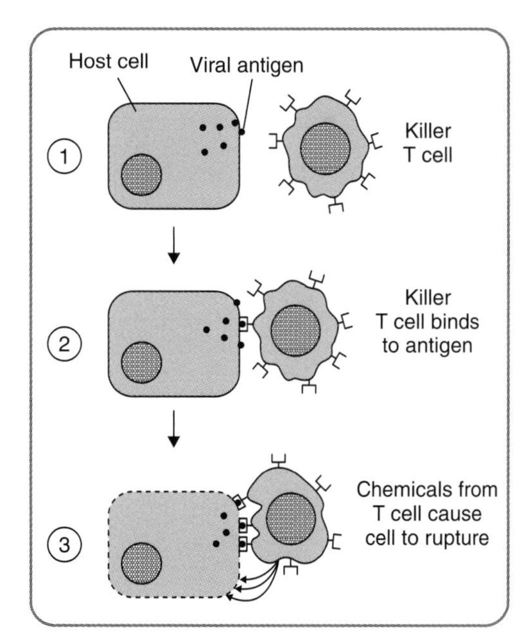

Figure 6.4 Stages in the cell-mediated response

Stage 1 When a cell is infected with a virus, 'foreign' protein released into the host cell is moved to the cell surface where it acts as a specific antigen.

Stage 2 A specific killer T cell recognises and binds to the antigens on the cell surface.

Stage 3 Chemicals produced by the killer T cell cause the cell to rupture.

Killer T cells also attack and destroy cancerous cells and cells of transplanted organs. In organ transplants this leads to tissue rejection. Tissue rejection is prevented by the administration of suppressor drugs that suppress the immune system.

Another type of T lymphocyte is the **helper T cell**. Helper T cells activate other immune cells including both killer T cells and B cells. Without helper T cells the immune system would not function.

Hints and Tips

You should know that it is the shape of the protein of the antibody that makes it specific to a specific antigen. (This is similar to the active site of an enzyme.)

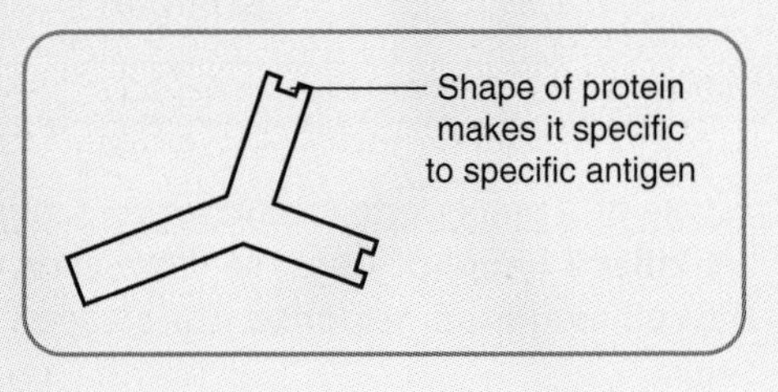

Figure 6.5 Shape of protein of antibody

T lymphocytes are so named due to the fact that after production from stem cells in the bone marrow, they have to be processed within the **thymus** gland. Both B and T lymphocytes mature and are activated within lymph nodes.

The HIV virus attacks helper T lymphocytes and as a result the immune system does not function. Some killer T cells survive as memory cells.

In a second infection by the same antigen the second response (secondary) is always faster than the first response (primary) due to the presence of both B and T memory cells. These replicate rapidly to re-establish protection.

For Practice

Using information from the text and Figure 6.4, copy and complete the flowchart to show the cell-mediated response.

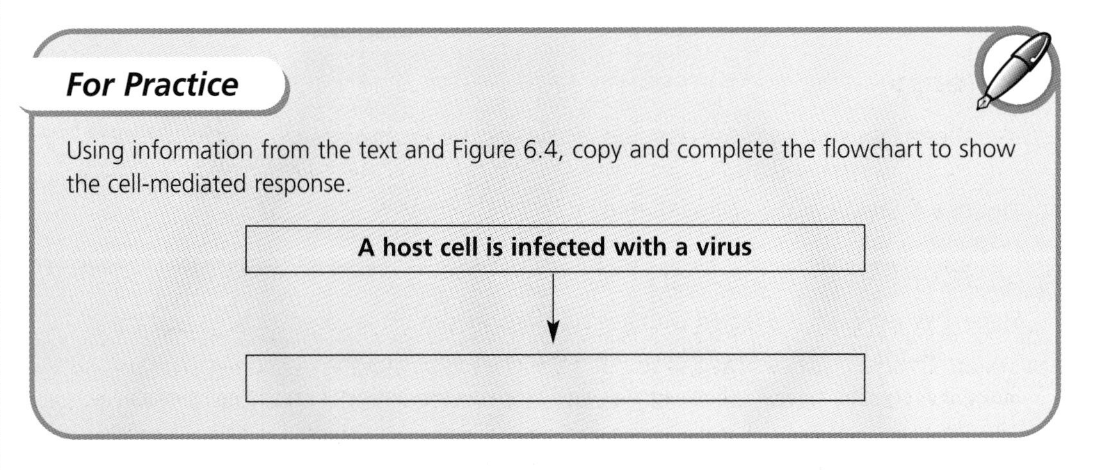

Immunity

Immunity is described as being innate or acquired. **Innate immunity** is inborn and includes our first and second lines of defence against viruses and bacteria. The first line of defence includes the skin and secretions passed onto the body surface. These include mucus and stomach acid. The second line of defence includes phagocytosis by macrophages.

Acquired immunity refers to the ways by which an individual becomes immune to antigens. Acquired immunity is obtained both naturally and artificially and it can be active or passive. The table below summarises these different forms of acquired immunity.

Type of acquired immunity	Definition	Further information
Natural – active	Antibodies are produced in response to foreign antigens.	This is the third line of defence of the body.
Natural – passive	Antibodies are received from the mother.	Antibodies pass across the placenta and are present in the mother's milk.
Artificial – active	Antibodies or antitoxins are produced after a vaccination containing antigens is given.	For example, if a vaccine containing inactivated toxin from the tetanus bacterium is given, then specific antibodies to the toxin are produced.
Artificial – passive	Antibodies or antitoxins to a disease are given through an inoculation.	For example, tetanus antitoxin is extracted from the pooled plasma of blood donors and injected into non-vaccinated individuals with deep and dirty wounds to prevent lockjaw.

Note: *Clostridium tetani* is a bacterium that causes tetanus in humans. Tetanus is a condition also referred to as lockjaw. If the bacteria enter through a wound, they produce a toxin that affects the nerves that control muscle activity.

Allergy

An **allergy** is an exaggerated response by the body to environmental antigens that normally would be considered to be harmless. Allergic responses that are common include those to pollen grains, peanuts, dust mites, bee venom and penicillin.

ABO blood grouping

The surface antigens on the red blood cells (RBCs) of an individual determine their blood group. There are two antigens, A and B. Individuals can possess one antigen and they are either group A or B; with both antigens they are group AB; and with neither they are group O. The antibodies in the blood plasma of an individual are determined by their blood group. Individuals with group A have anti-B antibodies; those with group B have anti-A; those with group AB have neither anti-A nor anti-B; and those with group O have both anti-A and anti-B.

Knowledge of an individual's blood group is essential for blood transfusion. Antibodies in the plasma of the recipient would attack 'foreign' antigens of an unsuitable donor.

For Practice

Copy and complete the table below to show which blood group(s) can be used in transfusions for the different blood groups. Group A has been completed for you.

Blood group	Surface antigens on RBC	Antibodies in plasma	Possible donors	Reason
A	A	Anti-B	A and O	Antibody B would not attack groups A or O but would attack B or AB
B				

Hints and Tips

Link vaccination to the section on 'Disease' in 'Population limiting factors', Chapter 16, page 161.

For use in vaccinations, disease-causing organisms are treated to make them less harmful.

Surface antigens determine the blood group. The antibodies present in the plasma are the ones opposite to the blood group.

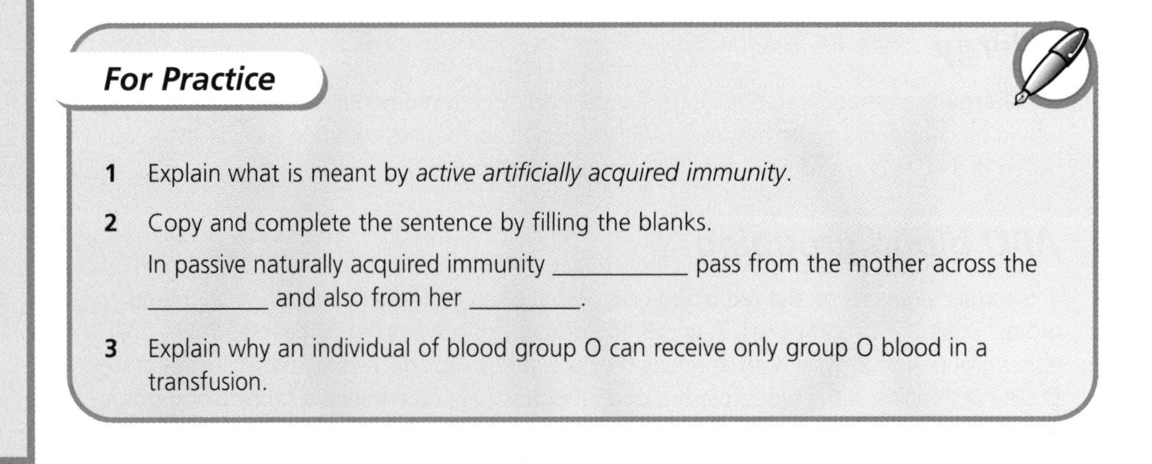

For Practice

1 Explain what is meant by *active artificially acquired immunity*.

2 Copy and complete the sentence by filling the blanks.

In passive naturally acquired immunity ___________ pass from the mother across the __________ and also from her _________.

3 Explain why an individual of blood group O can receive only group O blood in a transfusion.

The nature of viruses

Viruses are distinct from other microscopic disease-causing organisms such as bacteria, protozoa and fungi. Viruses are described as being living only in so far as they carry out reproduction of new viruses. Even this process requires the assistance of a cell which is damaged as a result. The **structures present in a virus** are shown in Figure 6.6.

The virus has an outer protein coat that surrounds a nucleic acid core. The nucleic acid is either DNA or RNA. Viruses are very small and range in size from about 10 to 100 nanometres (1 nanometre is 10^{-9} of a metre).

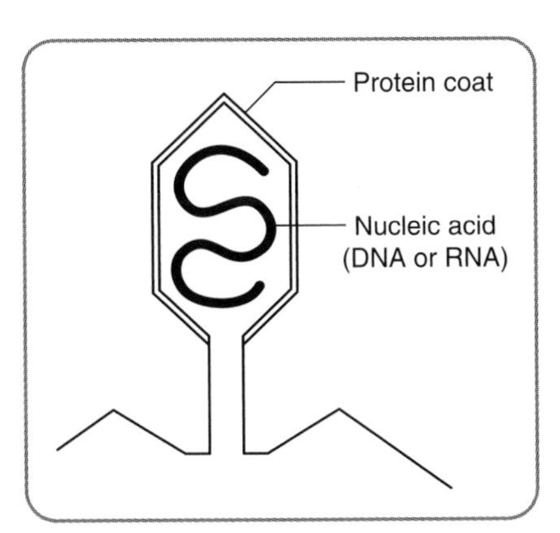

Figure 6.6 Structures in a virus

Viral replication

The stages in viral replication are shown in Figure 6.7.

Stage 1 The virus attaches to the surface of a specific host cell.

Stage 2 The viral nucleic acid is passed into the cell. The viral nucleic acid alters the metabolism of the cell and takes over control of the host cell.

Stage 3 The viral nucleic acid is replicated using the nucleotides and ATP of the host cell.

Stage 4 The nucleic acid of the viruses controls synthesis of new protein coats using the amino acids, ribosomes and ATP of the host cell.

Stage 5 New viruses are assembled as the nucleic acid enters a protein coat.

Stage 6 The new viruses are released from the host cell when the cell bursts.

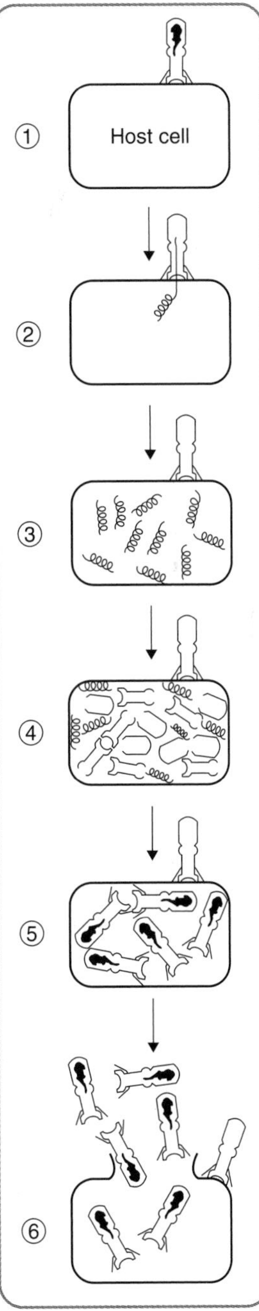

Figure 6.7 Stages in viral replication

Hints and Tips

The bursting of the cell to release new viruses is called cell lysis.

Each type of virus is specific to the cells of a particular host.

For Practice

1 Complete the blanks in the following sentences.
Viruses are classified as living due to the fact that they can _____________.
Viruses vary in size from around _____ to ____ nanometres.
Each virus consists of a ___________ coat that surrounds a ___________ core.

2 Use the information in the text to label a copy of Figure 6.7. For example, at Stage 1 add the label 'The virus attaches to the surface of the host cell'.

3 Using the information from the text and Figure 6.7, copy and complete the flowchart to show viral replication. The first box has been completed for you.

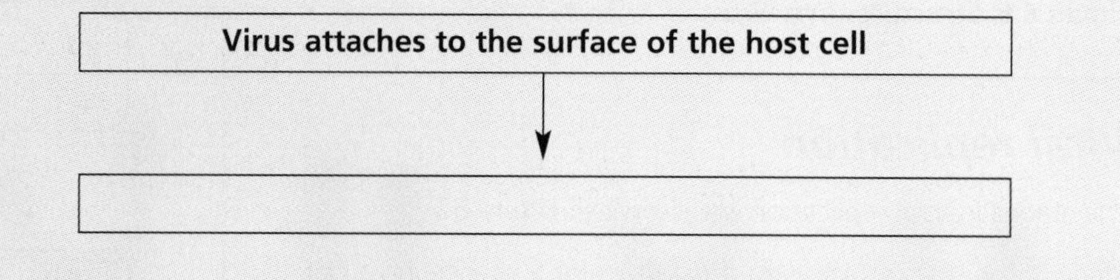

4 Make flashcards for all of the words that appear in **bold** type in the text of this chapter.

Exam Questions

1 Figure 6.8 shows the release of molecule X into a bacterial cell by a virus.

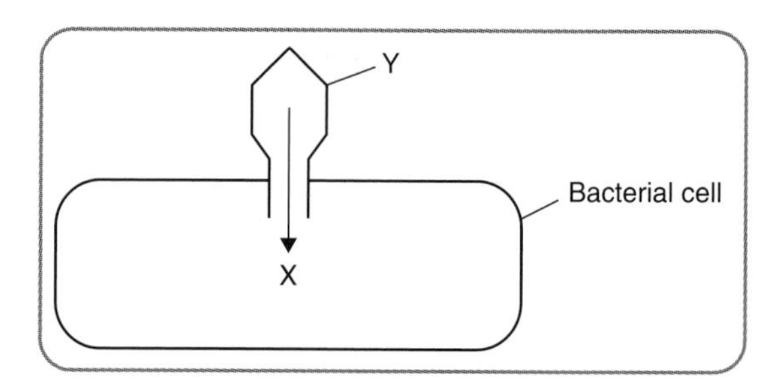

Figure 6.8 Release of molecule into bacterial cell

Exam Questions continued

Which line in the table below identifies structure Y, molecule X and the role of molecule X in viral replication?

	Structure Y	Molecule X	Role of molecule X in viral replication
A	Protein coat	Enzyme	Alteration of cell metabolism
B	Cellulose coat	DNA	Synthesis of new viral coats
C	Protein coat	RNA	Alteration of cell metabolism
D	Cellulose coat	Enzyme	Synthesis of new viral coats

2 An individual is blood group O. Which line in the table below identifies the antigens and antibodies present in their blood?

	Antigen on red blood cells	Antibodies in plasma
A	A and B	Anti-A and anti-B
B	None	Anti-A and anti-B
C	A and B	None
D	None	None

3 The flowchart in Figure 6.9 relates to the role of lymphocytes in defence.

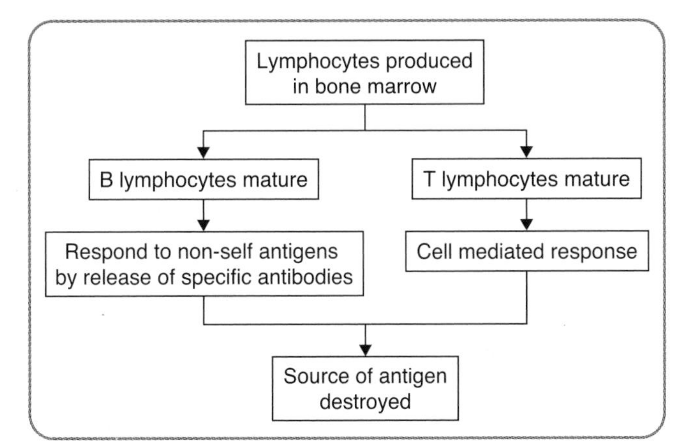

Figure 6.9 Flowchart: role of lymphocytes in defence

(a) What term is used to describe the response of B lymphocytes to antigens? *(1)*

(b) Explain the terms *specific antibodies* and *non-self antigens*. *(2)*

(c) Describe the cell mediated response of T lymphocytes. *(1)*

(d) Which type of immunity is acquired by the response of B lymphocytes to a virus? Explain your answer. *(2)*

4 Figure 6.10 represents the body's response to the BCG vaccination against the bacterium that causes tuberculosis.

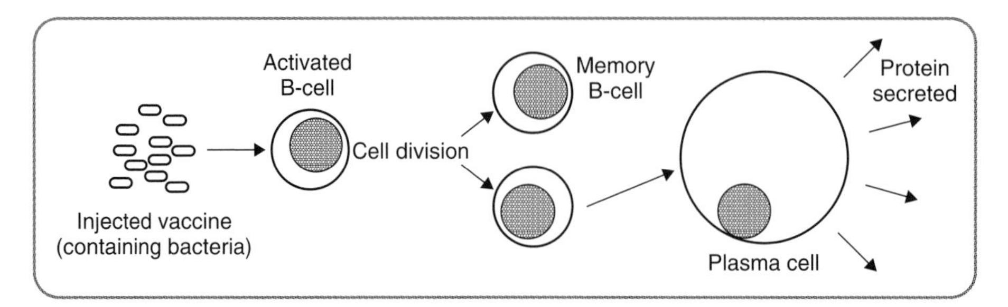

Figure 6.10 Response to BCG vaccine

(a) What feature of the bacterium activates the B-cells? *(1)*
(b) Name the type of protein secreted and state its function. *(1)*
(c) Suggest the function of the memory B-cells in the immune response. *(1)*
(d) Why do the bacteria present in the vaccine not cause tuberculosis? *(1)*

Answers

1 You should know that Y is the protein coat, X is the nucleic acid (DNA or RNA) and its role is to alter cell metabolism. **Answer** C

2 You should know that surface antigens determine the blood group and the antibodies present in the plasma are the ones opposite to the blood group. Group O has no antigens and therefore has anti-A and anti-B. **Answer** B

3 (a) Secretion of antibodies = Humoral response.
(b) Specific antibodies – proteins that are active against only one type of antigen. Non-self antigens – proteins on the surface of foreign cells such as invading organisms or a transplanted organ.
(c) They destroy infected cells.
(d) This is natural active immunity as the antibodies were produced in response to the presence of viral antigens.

4 (a) You should know that activation is in response to surface antigens.
(b) Protein = antibodies. They lead to destruction of the bacteria.
(c) You should remember that if the same antigen enters the body again the memory cells can quickly produce more of the cells to produce antibodies specific to that antigen.
(d) This is the idea that bacteria used in a vaccine are made harmless.

Chapter 7

CHROMOSOMES AS VEHICLES OF INHERITANCE

Key Ideas

Inheritance is the study of how genes are passed from one generation to the next.

Importance of DNA replication

Genes are specific regions of chromosomal DNA that determine characteristics of an organism. Figure 7.1 shows the location of some genes present on human chromosome 11. In cell division all cell structures must be replicated so that each new (daughter) cell receives a supply of all the structures required for survival. It is of particular importance for cell survival that accurate replication of all chromosomal DNA takes place.

The important facts of DNA replication include:

- Two exact copies of the genetic information are produced.
- In cell division, the nucleus divides in such a way that each daughter cell receives an exact copy of all the genetic information.
- If DNA replication goes wrong it can lead to mutations and cancers.

Part of human chromosome 11

- Catalase
- Calcitonin (hormone)
- Parathyroid hormone
- β-globulin
- Insulin growth factor 2

Figure 7.1 Location of genes on part of chromosome 11

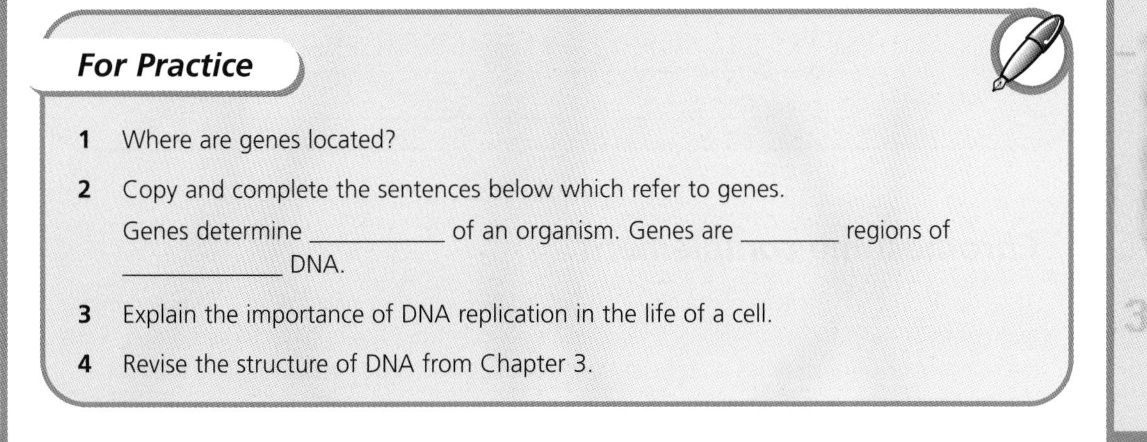

For Practice

1 Where are genes located?

2 Copy and complete the sentences below which refer to genes.

Genes determine __________ of an organism. Genes are ________ regions of ____________ DNA.

3 Explain the importance of DNA replication in the life of a cell.

4 Revise the structure of DNA from Chapter 3.

DNA replication

Before DNA replication takes place the appropriate enzymes, ATP and free DNA nucleotides must be present. The stages in DNA replication are outlined in Figure 7.2.

Stage 1 The DNA double helix unwinds.

Stage 2 Hydrogen bonds between complementary base pairs break and DNA unzips. Bases of strands are exposed.

Stage 3 Free DNA nucleotides bond with complementary exposed bases. Base pairings are A to T, T to A, C to G and G to C.

Stage 4 Adjoining nucleotides bond through sugar of one to phosphate of an adjacent nucleotide.

Stage 5 Two identical daughter DNA strands are formed. Each molecule contains one of the original strands.

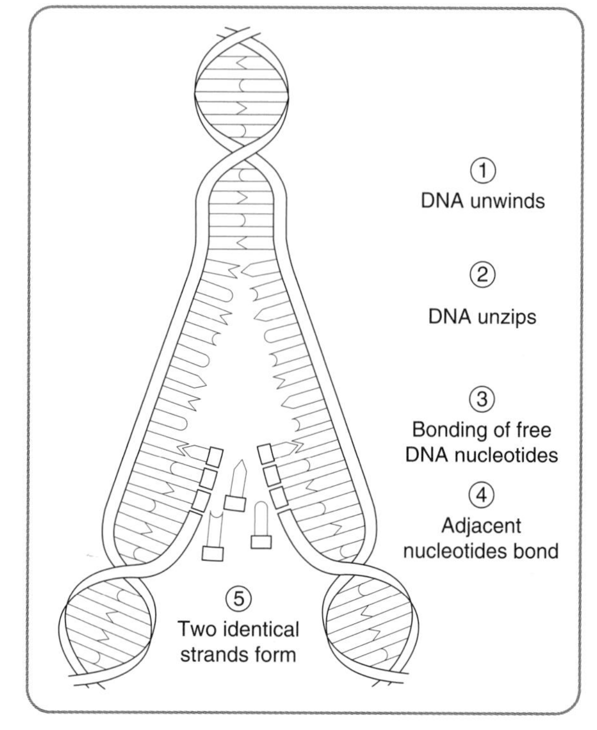

Figure 7.2 DNA replication

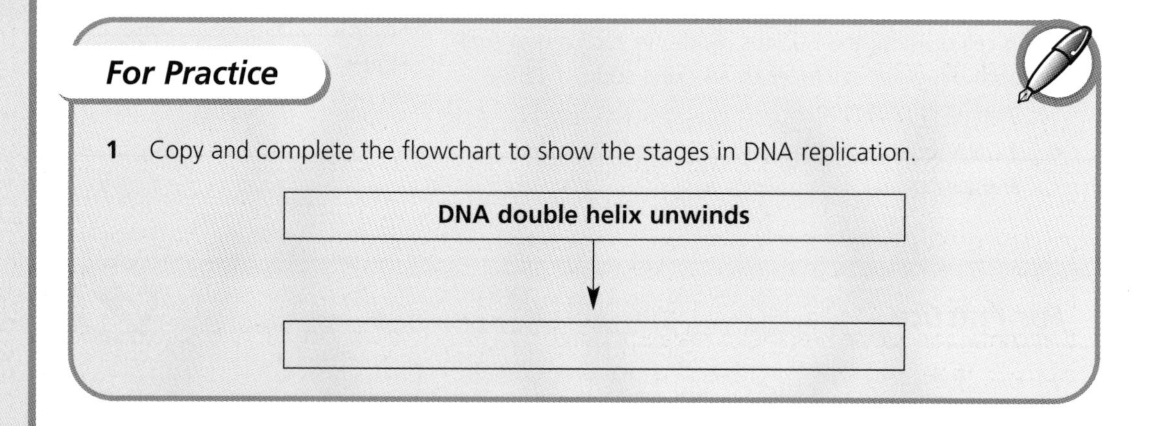

For Practice

1 Copy and complete the flowchart to show the stages in DNA replication.

Chromosome complement

At fertilisation each gamete contains 23 chromosomes and they give rise to a fertilised egg that contains 46 chromosomes. Each body cell develops from the original fertilised egg. Thus the number of chromosomes present in the nucleus of each body cell is 46. This number is the **chromosome complement**.

Figure 7.3 shows the arrangement of the 46 chromosomes that make up the human chromosome complement into their homologous pairs. This is called a **karyotype**. The karyotype of a female shows the chromosomes set out in matching pairs. The chromosomes match in size, in position of the centromere and in the location of genes on the chromosome; however, the alleles of the genes may differ. The pair shown as XX are the **sex chromosomes** and these determine the sex of the individual. The other 22 matching pairs are referred to as **autosomes**. These play no part in sex determination. The matching pairs of chromosomes are referred to as **homologous pairs**.

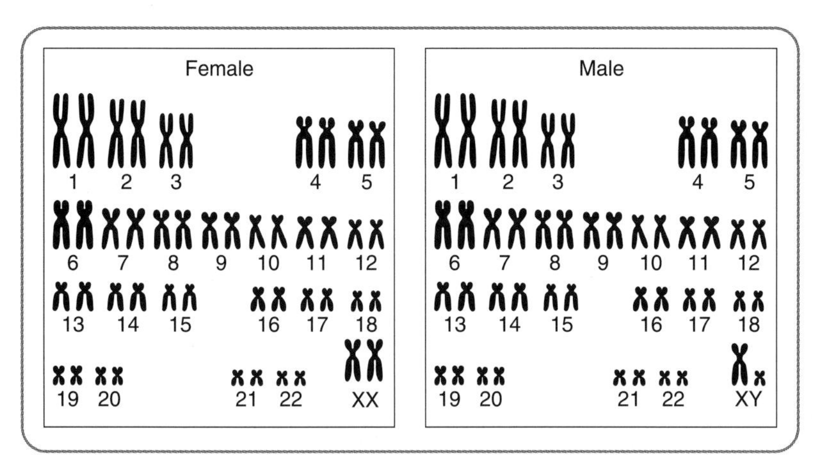

Figure 7.3 Arrangement of human chromosomes

From the karyotype of a male it can be seen that the 44 autosomal chromosomes form 22 homologous pairs but that the sex chromosomes differ. One is an X-chromosome and the other is a Y-chromosome. These chromosomes are only partially homologous.

Sex determination

A pair of chromosomes determines sex in humans. Two X chromosomes determines a female. One X and one Y chromosome determines a male. Figure 7.4 shows how the sex of an individual is inherited.

In gamete formation the sex chromosomes align and separate in the same way as other chromosomes. A female is XX and produces only **female gametes** that contain an X chromosome. A male is XY and produces **male gametes** that contain either an X or a Y chromosome.

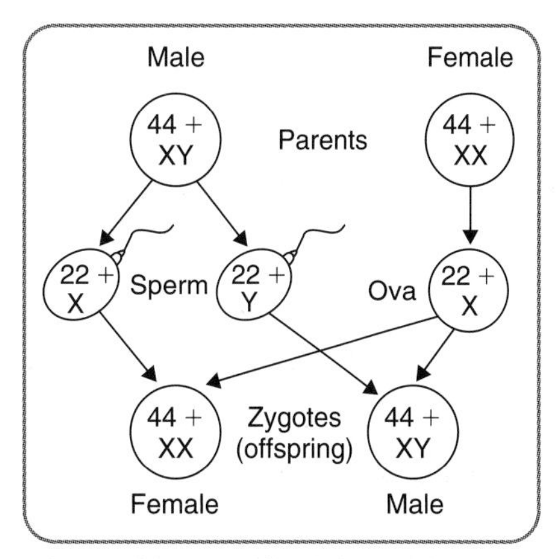

Figure 7.4 Sex inheritance

Outline of meiosis

Meiosis is a cell division in which the number of chromosomes in a **gamete mother cell** is reduced from two sets (**diploid**) to a single set (**haploid**). Meiosis occurs only in specialised body cells in the testes and ovaries.

In Figure 7.5, showing the stages in meiosis, shaded chromosomes were donated originally by one of the parents in the ovum and unshaded chromosomes by the other parent in the sperm. Only two pairs of the homologous chromosomes are shown.

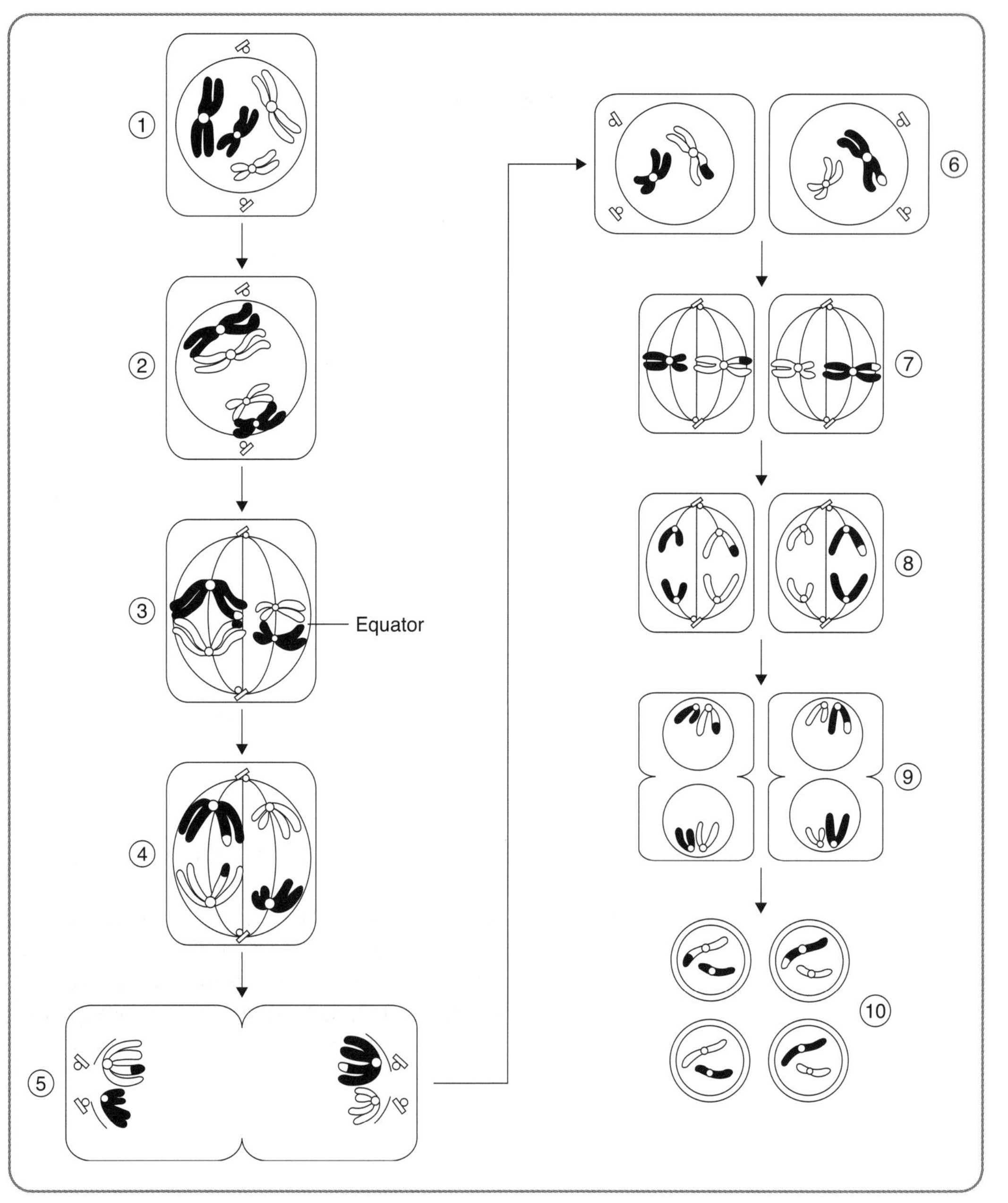

Figure 7.5 Meiosis

First meiotic division

Stage 1 At the start of meiosis, as a result of DNA replication, each chromosome exists as two genetically identical chromatids attached at the centromere.

Stage 2 Chromosomes align in homologous pairs. Points of contact called **chiasmata** form between the chromatids of members of the same homologous pairs.

Stage 3 Chromosome material breaks off from both chromatids at the chiasmata and crosses over. The homologous pairs move to the equator of the cell. Homologous pairs align at random (only one of the two possible arrangements is shown here – see Figure 7.6).

Stage 4 The homologous chromosomes are pulled to opposite poles.

Stage 5 The homologous pairs are separated. The cell membrane starts to pinch inwards.

Stage 6 The cell divides to form two cells, each with a haploid set of chromosomes.

Second meiotic division

Stage 7 The chromosomes align at the equator of the cell.

Stage 8 The chromatids in each cell are pulled towards opposite poles of the cells.

Stage 9 The cell membrane pinches inwards and the cells start to divide. New nuclear membranes are formed.

Stage 10 As a result of cell division four cells are formed. Each cell has the haploid chromosome number.

For Practice

1 Use information from the text to label a copy of Figure 7.5. For example, at Stage 1 label the chromatids and centromeres.

2 Using the information from the text and Figure 7.4, copy and complete the flowchart to show the first stage of meiosis. Two boxes have been completed for you.

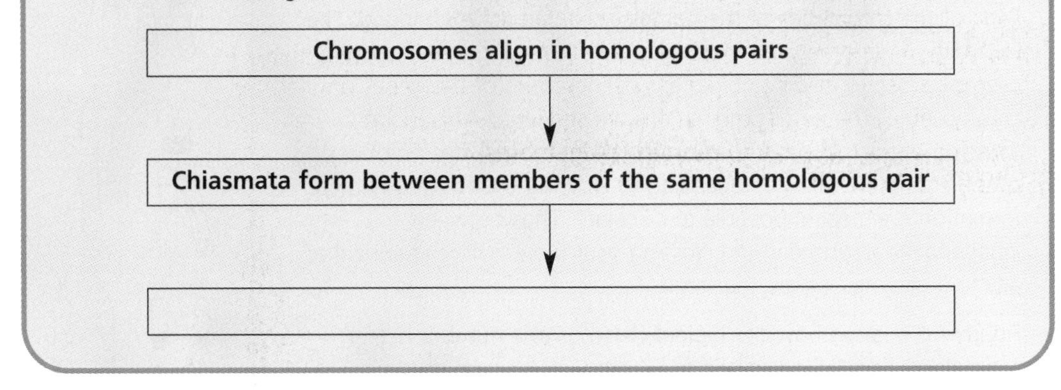

Independent assortment

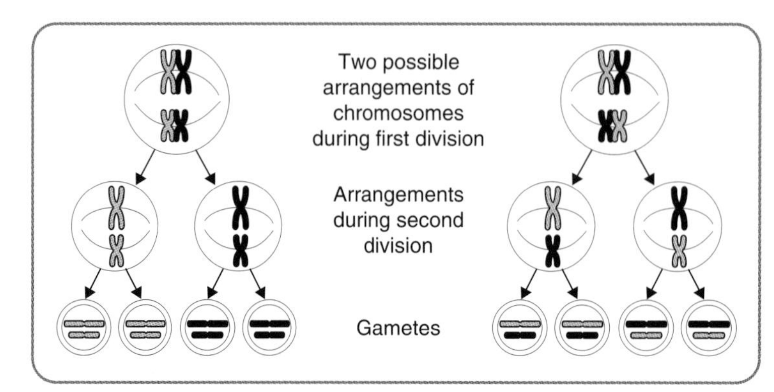

Figure 7.6 Alignment of homologous chromosomes

Figure 7.6 shows possible alignments of two pairs of homologous chromosomes during the first meiotic division. Shaded chromosomes were donated originally by one of the parents in the ovum and unshaded chromosomes by the other parent in the sperm.

The random alignment of the homologous pairs is called **independent assortment**. As a result of independent assortment, four types of gamete are formed. Independent assortment of the homologous pairs ensures variation in the gametes. In humans with 23 homologous pairs, these can align in 2^{23} (more than 8 million) different ways.

Crossing-over

Figure 7.7 shows the result of **crossing-over** between members of a homologous pair during meiosis. The position of the genes, together with their possible alleles, are shown as A, a, B, b, C and c. A chiasma is formed between genes A and B.

Parts of the chromatids of the homologous pair are exchanged. The exchange of chromosome material is called crossing-over. Crossing-over gives rise to new combinations of the alleles of the genes on the chromosome. The new combinations are shown as Abc and aBC. Crossing-over produces **variation in the gametes**. The number of possible different chiasmata leading to crossing-over during gamete formation is almost impossible to calculate. Thus between independent assortment and crossing-over there is little chance that any two gametes will be the same.

Figure 7.8 outlines how the diploid chromosome number is maintained. Specialised cells in the reproductive tissue divide by meiosis to form haploid gametes. Due to the resulting variation from independent assortment and crossing-over, each gamete has a unique genotype. **Fertilisation** returns the chromosome number to the diploid number. With each gamete being unique, each fertilised egg will show a unique genotype. As a result of growth and development the fertilised egg, over time, develops into an adult human.

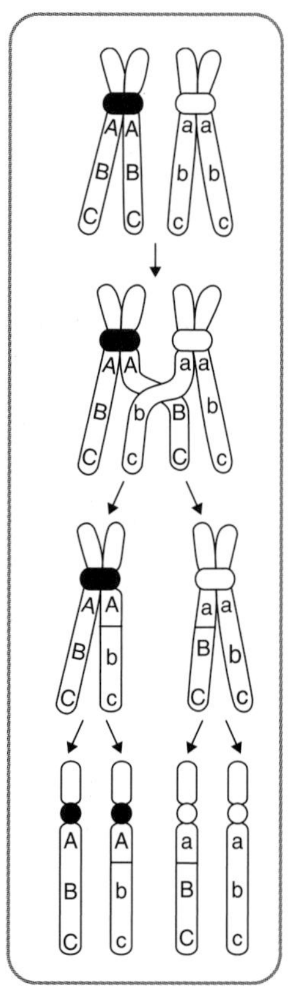

Figure 7.7 Crossing-over

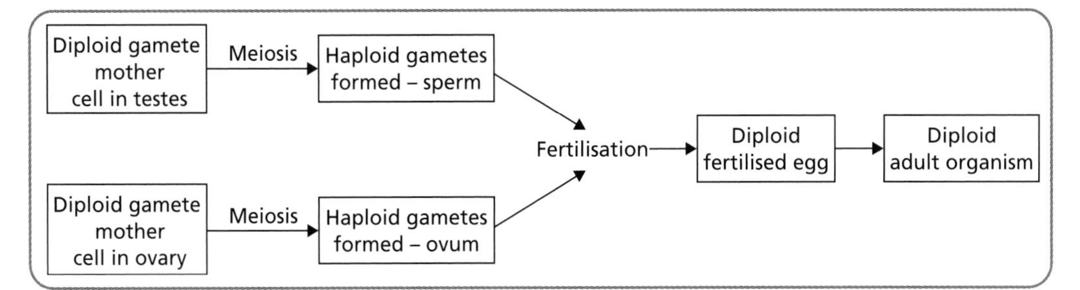

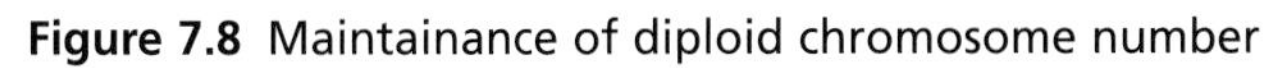

Figure 7.8 Maintainance of diploid chromosome number

Hints and Tips

In the first meiotic division homologous chromosomes are separated.

In the second meiotic division chromatids are separated.

The haploid chromosome number is represented by n where n = 1 set of chromosomes.

The diploid chromosome number is represented by 2n.

For Practice

1. Show the possible alignment of three pairs of homologous chromosomes during the first meiotic division.
2. Use information from the text to label the chromosomes in Figure 7.7. For example, draw a label with an arrow pointing to '*Alleles of a gene*'.
3. Name two sources of variation in the process of meiosis.
4. Name the structures that are separated during the first meiotic division.
5. Name the structures that are separated during the second meiotic division.
6. Make flashcards for all of the words that appear in **bold** type in the text of this chapter.

Exam Questions

1 Which of the following shows the chromosome number and the genetic composition of cells that are produced by meiosis?

	Chromosome number	Genetic composition
A	Diploid	Cells show variation
B	Diploid	Cells are identical
C	Haploid	Cells show variation
D	Haploid	Cells are identical

2 Figure 7.9 represents an early stage in the process of meiosis in the ovary of an animal cell.

Which letter shows the homologous partner of chromosome P?

A Q
B R
C S
D T

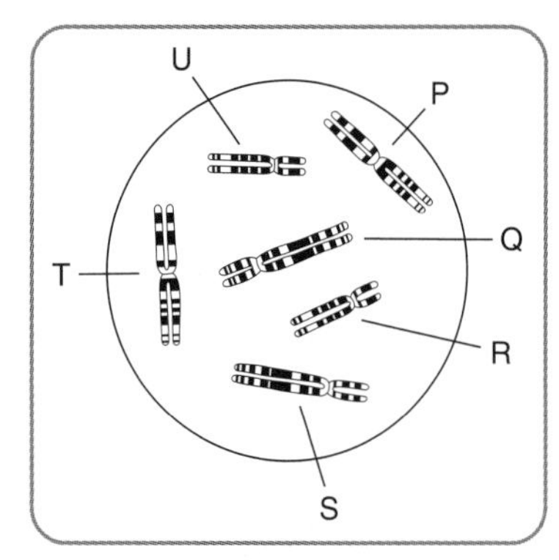

Figure 7.9 Early meiosis

3 Crossing-over is a process that involves the:

A exchange of genetic material between non-homologous chromosomes
B exchange of genetic material between homologous chromosomes
C random alignment of homologous chromosomes
D random alignment of non-homologous chromosomes.

4 Figure 7.10 shows the stained chromosome complement from the cell of an individual.

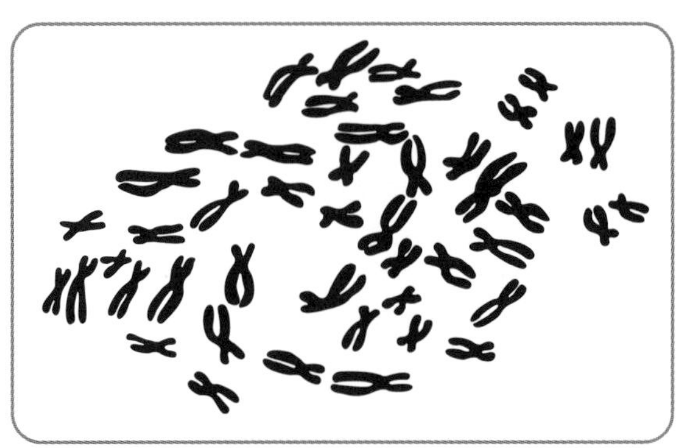

Figure 7.10 Chromosome complement

Exam Questions continued

(a) To examine the chromosome complement, the chromosomes are cut out and arranged in homologous pairs. State **two** features that allow homologous pairs of chromosomes to be identified. *(1)*
(b) What name describes the arrangement of the 46 chromosomes that make up the human chromosome complement into their homologous pairs? *(1)*
(c) How could the sex of the individual be identified from this paired arrangement? *(1)*

5 Figure 7.11 shows stages in the process of meiosis. The diagrams are not in the correct order and only six chromosomes are shown.

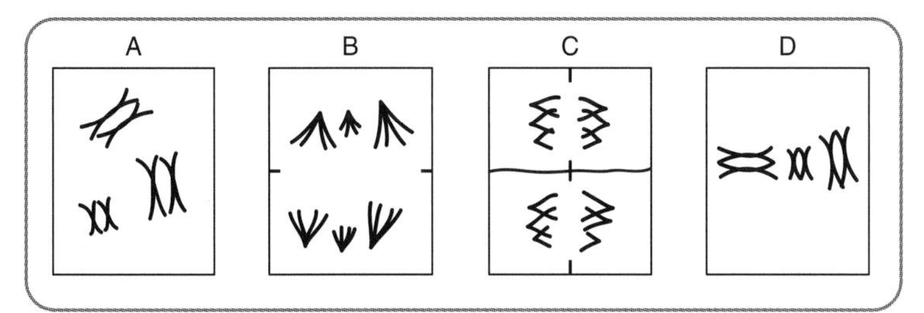

Figure 7.11 Stages of meiosis

(a) (i) Place the stages in the correct order. *(1)*
(ii) At which stages could chiasmata form? *(1)*
(iii) State the importance of chiasmata. *(1)*
(b) Figure 7.12 shows a gamete mother cell and four sperm cells that result from meiosis.
(i) Complete the diagram by writing in the normal number of chromosomes present in each of the cells. *(1)*

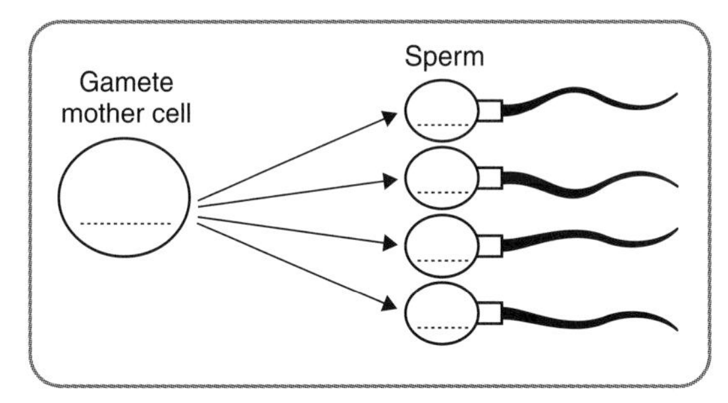

Figure 7.12 Gamete mother cell and sperm

(ii) In how many of these sperm will an X chromosome be present? *(1)*

Answers

1 Cells produced by meiosis are gametes and they are haploid and show variation due to independent assortment and crossing-over. **Answer** C

2 Examine the position of the centromere, the length and banding of the chromosomes to find the identical chromosome. **Answer** D

3 You should know that crossing-over is the exchange of chromatid material between members of a homologous pair. **Answer** B

4 (a) Features are: same position of the centromere; the same length.
(b) You should know that this is the karyotype.
(c) Sex is determined by sex chromosomes.

Answer If both sex chromosomes were X then it is a female; if one chromosome is an X and the other a Y then it is a male.

5 (a) (i) In A homologous chromosomes are pairing. In B homologous pairs of chromosomes are being separated. In C chromatids are being separated. In D homologous pairs are aligned at the equator. **Answer** A→D→B→C
(ii) You should know that chiasmata form as the homologous chromosomes pair. This is shown in diagram **A** and **D**.
(iii) You should know that chiasmata may lead to crossing-over and that this produces variation in the gametes.

Answer Chiasmata may lead to crossing-over and produces variation in the gametes.

(b) (i) You should know that the human gamete mother cell is diploid = 2 sets = 46 chromosomes and that gametes (sperm) are haploid = 1 set = 23.
(ii) You should know that a male has X and Y sex chromosomes. Chromosomes replicate; X and Y separated at first meiotic division; chromatids of X separated at second meiotic division. **Answer** 2

Chapter 8

MONOHYBRID INHERITANCE, MUTATION AND CHROMOSOME ABNORMALITIES

Key Ideas

Monohybrid inheritance is the pattern of inheritance of a pair of alleles where one is dominant and one is recessive. The list of terms used in genetic crosses includes:

Haploid – the one set chromosome number (n)

Diploid – the two set chromosome number (2n)

Allele – different forms of a gene

Multiple alleles – a gene with three or more alleles

Genotype – the genetic make-up of an individual

Phenotype – the appearance of an individual

Dominant – the allele which is expressed in an individual with two different alleles

Co-dominant – both alleles are expressed in an individual

Incomplete dominance – the recessive allele has a partial effect on the phenotype

Recessive – the allele which is not expressed in an individual with two different alleles

Homozygous – an individual having an identical pair of alleles of a gene

Heterozygous – an individual having two different alleles of a gene

Carrier – a heterozygote who may pass the recessive allele on to the next generation

Cross – interbreeding of two individuals to produce offspring

P – identifies the parents in a cross

F1 – identifies the offspring in the first generation in a cross

F2 – identifies the offspring in the second generation in a cross

Monohybrid cross – a cross in which the inheritance of one pair of alternative characteristics is followed, e.g. pigmented skin and albino

Nomenclature

Letters are used to identify the alleles of genes. Dominant alleles are represented by the uppercase of the letter and recessive alleles are represented by the lowercase of the letter, for example the allele for tongue rolling is dominant to that for non rolling. Then R for roller and r for non roller. In co-dominance both alleles are represented by the uppercase of the letter.

Hints and Tips

Remember that each individual has two alleles of each gene in their body cells.

As you work your way through the cross, check on the list of definitions to reinforce your understanding of the cross. For example, if an individual is homozygous the two alleles are the same.

In gamete formation (meiosis) the matching pairs of chromosomes are separated, so there is only one allele of each gene in a gamete.

In a heterozygote, as a result of separation of homologous pairs of chromosomes during meiosis, gametes R and r are produced in equal numbers.

Gametes combine at random – to ensure that you get the combinations correct, use a Punnett square.

In crosses the expected ratios may not be obtained for the following reasons:

1 Sample size is too small

2 Fertilisation is a random process.

Example of a monohybrid cross to the F_2 generation

In humans the gene for tongue rolling has two alleles. The allele for a roller (R) is dominant to the allele for a non roller (r). A homozygous female roller marries a homozygous male non roller and they have a son and a daughter. This cross is shown below.

P	Phenotypes	(Female) roller	X	(Male) non roller
	Genotypes	RR	X	rr
	Gametes	R		r
F1	Genotype of children	Rr		
F1	Phenotype of children	All rollers		

The daughter from the F1 marries a man heterozygous for the condition. The possible phenotypes and genotypes of their children are shown on the following page.

***Example** continued* ➢

***Example** continued*

Phenotypes	(Female) roller	X	(Male) roller
Genotypes	Rr	X	Rr
Gametes	R *or* r		R *or* r

Punnett square

Gametes	R	r
R	RR	Rr
r	Rr	rr

From the results:

Expected F2 genotype ratio is 1 RR : 2 Rr : 1 rr

Expected **F2** phenotype ratio is 3 rollers : 1 non roller

For Practice

In humans the gene for a red blood cell (RBC) antigen has two alleles. The allele for the presence of antigen D is dominant to the allele for the absence of antigen D(d). A female homozygous for the presence of antigen D marries a male homozygous for the absence of antigen D and they have a son and a daughter.

The son from the F1 marries a woman heterozygous for the condition. Outline the above information in a monohybrid cross to the F2 generation as shown in the example.

Co-dominance

With co-dominant alleles, both alleles are expressed in the phenotype of a heterozygous individual. Antigens M and N are proteins on the surface of RBCs that show **co-dominance**. Individuals with genotype MM have blood group M; individuals with NN have blood group N; and individuals with MN have blood group MN. In co-dominance a different uppercase letter represents each allele.

Example of a co-dominance cross

A woman with blood group M marries a man with blood group N and they have a son. The son later marries a woman with blood group MN. The cross is outlined below to the F2 generation.

P	Phenotypes	(Female) blood group M	X	(Male) blood group N
	Genotypes	MM	X	NN
	Gametes	M		N
F1	Genotype of son	MN		
F1	Phenotype of son	Blood group MN		

The son marries a woman with blood group MN. The possible phenotypes and genotypes of their children are shown in the cross below.

Phenotypes	(Female) blood group MN	X	(Male) blood group MN
Genotypes	MN	X	MN
Gametes	M *or* N		M *or* N

Punnett square

Gametes	M	N
M	MM	MN
N	MN	NN

From the results:

Expected F2 genotype ratio is 1 MN : 2 MN : 1 NN

Expected F2 phenotype ratio is 1 group MN : 2 group MN : 1 group NN

Incomplete dominance

In **incomplete dominance** the recessive allele has a partial effect on the phenotype. Acute sickle-cell anaemia is a recessive condition in which abnormal haemoglobin is synthesised. A homozygous recessive individual suffers from

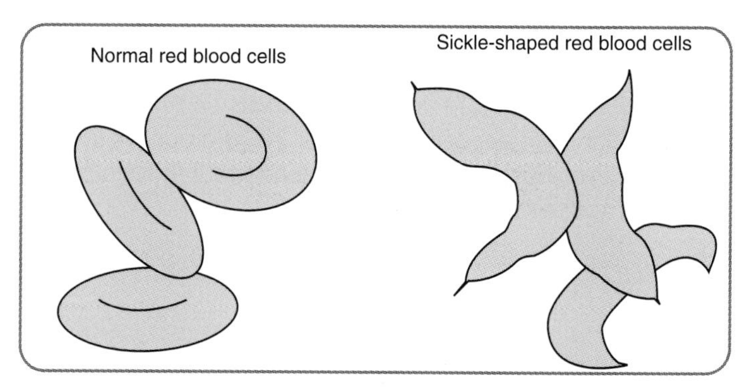

Figure 8.1 Normal and sickle-shaped red blood cells

acute sickle-cell anaemia. The RBCs collapse and stick together and oxygen uptake and transport are reduced (Figure 8.1). Such individuals generally die at a young age. A homozygous dominant individual has the normal healthy condition. A heterozygous individual exhibits symptoms of the disorder and has the sickle-cell trait. The nomenclature for the condition is H representing the normal allele and H^S the recessive allele.

Example of incomplete dominance

Two individuals with the sickle-cell trait have a child. What is the chance that their child will suffer from sickle-cell anaemia?

The cross is shown below.

P	Phenotypes	(Female) sickle-cell trait	X	(Male) sickle-cell trait
	Genotypes	HH^S	X	HH^S
	Gametes	H *or* H^S		H *or* H^S

Punnett square

Gametes	H	H^S
H	HH	HH^S
H^S	HH^S	H^SH^S

From the results:

Expected F1 phenotype ratio is 1 normal : 2 sickle-cell trait : 1 sickle-cell anaemia. Of the four possible outcomes only one shows acute sickle-cell anaemia. Chance is 1 in 4.

Multiple alleles

A gene that has three or more alleles exhibits multiple alleles. The A B O blood group system (Chapter 6) is determined by three alleles of the same gene. Allele A produces antigen A; allele B produces antigen B; and allele O produces no antigen. Alleles A and B are co-dominant; allele O is recessive to both A and B. The table below shows the possible genotypes of each blood group.

Blood group	Alleles of possible genotypes
A	AO or AA
B	BO or BB
AB	AB
O	OO

Hints and Tips

An individual who has a homozygous recessive parent must receive one copy of the recessive allele from that parent.

An individual with blood group A is either homozygous (AA) or heterozygous (AO) and a blood group B individual is either BB or BO.

Example of multiple alleles

The possible phenotypes and genotypes of the children of parents, one of blood group A whose mother was group O and the other of blood group B whose mother was group O, are shown in the cross below.

P	Phenotypes	Blood group A	X	Blood group B
P	Genotypes	AO	X	BO
P	Gametes	A *or* O		B *or* O

Punnett square

Gametes	A	O
B	AB	BO
O	AO	OO

From the results:

Expected **genotype** ratio is 1 AB : 1 AO : 1 BO : 1 OO

Expected **phenotype** ratio is 1 Group AB : 1 Group A : 1 Group B : 1 Group O

For Practice

Show the possible offspring of a cross between a mother heterozygous for blood group B whose husband is blood group AB.

Sex-linked inheritance

Figure 8.2 shows the structure of the X and Y chromosomes. The Y chromosome is smaller. The Y chromosome has an area that is homologous with an area on the X chromosome. The section of the Y chromosome that is missing has no homologous area on the X chromosome. This area, therefore, lacks a section of genes. This area is called the non-homologous area of the X chromosome and the genes on this area are called **sex-linked** genes.

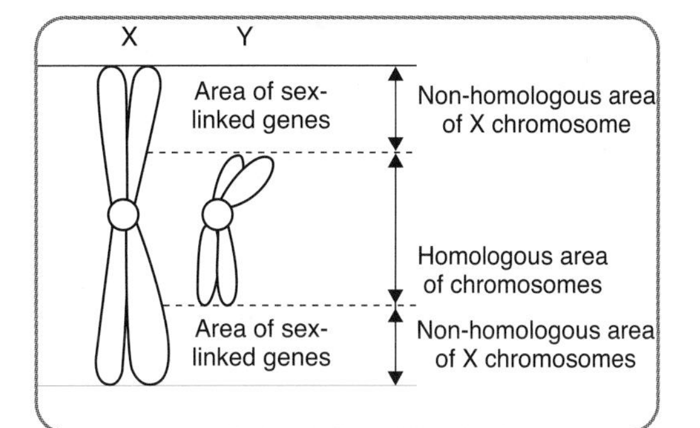

Figure 8.2 Sex chromosomes

Sex-linked crosses

Chromosomes are represented by the uppercase letters X and Y. The alleles are represented by the appropriate upper- and lowercase superscripts. For example: red–green colour deficiency is sex-linked. The normal allele (X^R) is dominant to the recessive allele (X^r) that causes colour deficiency. The phenotypes of the five possible genotypes are listed in the table below.

Phenotype	Genotype
Female with normal red–green colour vision (homozygous dominant)	$X^R X^R$
Female with normal red–green colour vision (heterozygous)	$X^R X^r$
Female with red–green colour deficiency (homozygous recessive)	$X^r X^r$
Male with normal red–green colour vision	$X^R Y$
Male with red–green colour deficiency	$X^r Y$

Females heterozygous for a sex-linked condition are referred to as carriers. They do not show the disorder but may pass it to the next generation.

Example of a sex-linked disorder

After a cut or wound the blood of a person with haemophilia takes a long time to clot. This disorder is sex-linked and is caused by a recessive allele. A carrier female and a normal male have children. What is the chance that any of the children will have haemophilia?

P	Phenotypes	Carrier female	X	Normal male
	Genotypes	$X^H X^h$	X	$X^H Y$
	Gametes	X^H or X^h		X^H or Y

Punnett square

Gametes	X^H	Y
X^H	$X^H X^H$	$X^H Y$
X^h	$X^H X^h$	$X^h Y$

***Example** continued* ➢

Example continued

Expected phenotype ratio: 1 normal : 1 carrier : 1 normal : 1 haemophiliac

female	female	male	male

The chance of a child having haemophilia may be viewed in two ways:

1 Of the four possible outcomes only one shows haemophilia.
Chance is 1 in 4.

2 In this cross it is only sons that inherit the disorder.
Chance is 1 in 2 that a son inherits the disorder.

Hints and Tips

A son inherits his X chromosome from his mother.

A daughter inherits her father's X chromosome.

Males are at a greater risk of inheriting sex-linked disorders as they require to have only one copy of the allele for the disorder to show.

A father can never pass a sex-linked disorder to his sons.

A grandfather can pass the disorder to his grandsons through his daughters.

For Practice

Red–green colour vision is sex-linked. Outline the cross between a woman who is heterozygous for red–green colour vision and her husband who has red–green colour deficiency, to show possible phenotypes and genotypes of their children.

Polygenic inheritance

In **polygenic inheritance**, two or more genes control the inheritance of a characteristic. These genes have two or more alleles. Characteristics such as height, skin colour and eye colour show polygenic inheritance.

Example of polygenic inheritance

The colour of the iris in the human eye is controlled by at least two genes. The dominant alleles of the genes (D and E) cause production of melanin in the iris (melanin is the pigment that colours skin and hair). The recessive alleles of the genes (d and e) do not produce melanin. The table below shows the range of colour production associated with different genotypes.

Number of alleles of D and E present	Possible genotypes	Colour of iris
4	DDEE	Dark brown
3	DDEe DdEE	Medium brown
2	DdEe DDee ddEE	Light brown
1	Ddee ddEe	Deep blue/Green
0	ddee	Light blue

A characteristic that shows a wide range of phenotypes shows **continuous variation**. Polygenic inheritance gives rise to continuous variation and this is shown in Figure 8.3 for height in a human population. The bell-shaped curve of the graph is referred to as the **normal distribution**. Most individuals are within a range that borders on either side of average height. Very few are either very short or very tall.

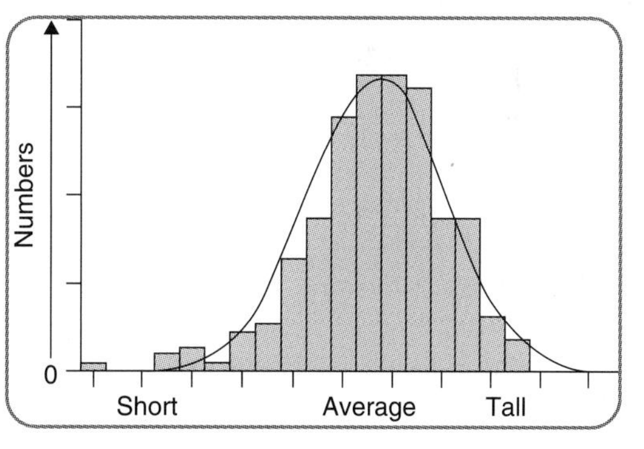

Figure 8.3 Continuous variation

Gene mutations

Look back at Chapter 3 to revise protein synthesis.

In a gene mutation:

- the order of bases in a DNA strand is altered
- the order of bases on the strand of mRNA synthesised by the DNA is altered
- a codon or codons are altered
- the order of attachment of tRNAs is altered
- the sequence of amino acids in the protein is altered
- the structure of the protein is changed.

Four types of gene mutation are shown in Figures 8.4–8.7, together with the consequent changes that occur in the mRNA and in the amino acid sequence in the protein. For simplification, only the strand of DNA that controls mRNA synthesis is shown in these figures.

Figure 8.4 shows a **substitution** mutation. One nucleotide pair is substituted with a different nucleotide pair. This alters the base sequence and thus changes a codon within the mRNA. In turn this alters only one amino acid in the protein and thus changes the structure of the protein formed slightly.

Normal DNA sequence	CTG	ACT	CCT	GAG	GAG
Normal mRNA sequence	GAC	UGA	GGA	CUC	CUC
Normal amino acid sequence	leucine	threonine	proline	glutamic acid	glutamic acid
Altered DNA sequence	CTG	ACT	CCT	GTG	GAG
Altered mRNA sequence	GAC	UGA	GGA	CAC	CUC
Altered amino acid sequence	leucine	threonine	proline	valine	glutamic acid

Figure 8.4 Substitution

Figure 8.5 shows an **inversion** mutation. Two or more nucleotide pairs break off, reverse and rejoin within the original area of the DNA. This reverses the base sequence and thus changes a codon or codons within the mRNA. This alters the amino acid sequence in the protein and thus changes the structure of the protein formed slightly.

Normal DNA sequence	GGT	CCT	CTC	ACG	CCA
Normal mRNA sequence	CCA	GGA	GAG	UGC	GGU
Normal amino acid sequence	proline	glycine	glutamic acid	cysteine	glycine
Altered DNA sequence	GGT	CTC	CTC	ACG	CCA
Altered mRNA sequence	CCA	GAG	GAG	UGC	GGU
Altered amino acid sequence	proline	glutamic acid	glutamic acid	cysteine	glycine

Figure 8.5 Inversion

Figure 8.6 shows a **deletion** mutation. One (or more) nucleotide pair breaks off and the DNA rejoins with this nucleotide pair missing. This alters the base sequence along the length of the DNA after the deletion and changes the codon sequence in the mRNA. This alters the amino acid sequence in the protein and thus changes the structure of the protein formed.

		Deletion ↓				
Normal DNA sequence	GGT	CCT	CTC	ACG	CCA	CCC
Normal mRNA sequence	CCA	GGA	GAG	UGC	GGU	GGG
Normal amino acid sequence	proline	glycine	glutamic acid	cysteine	glycine	etc.
Altered DNA sequence	GGT	CTC	TCA	CGC	CAC	
Altered mRNA sequence	CCA	GAG	AGU	GCG	GUG	
Altered amino acid sequence	proline	glutamic acid	cysteine	alanine	valine	

Figure 8.6 Deletion

Figure 8.7 shows an **insertion** mutation. One (or more) nucleotide pair is added into the DNA chain. This alters the base sequence along the length of the DNA after the insertion and changes the codon sequence in the mRNA. This in turn alters the amino acid sequence in the protein and thus changes the structure of the protein formed.

Point of insertion of nucleotide with base A (between C and T)

Normal DNA sequence	GGT	CCT	CTC	ACG	CCA	
Normal mRNA sequence	CCA	GGA	GAG	UGC	GGU	
Normal amino acid sequence	proline	glycine	glutamic acid	cysteine	glycine	etc.
Altered DNA sequence	GGT	CCA	TCT	CAC	GCC	A etc.
Altered mRNA sequence	CCA	GGU	AGA	GUG	CGG	U etc.
Altered amino acid sequence	proline	glycine	arginine	valine	arginine	

Figure 8.7 Insertion

Hints and Tips

For gene mutations remember **SIDI** – **S**ubstitution, **I**nversion, **D**eletion and **I**nsertion.

Gene mutations lead to changes in the structure of the protein formed.

If the structure of the active site of an enzyme is changed then the substrate may no longer fit to the active site.

Sickle-cell anaemia is caused by a substitution mutation.

Be able to link, in both directions: changes in the base sequence on DNA to the base sequence on mRNA to the specific tRNAs to the amino acid sequence.

For Practice

1. Highlight the base of the nucleotide that was substituted in Figure 8.4 and the change in the amino acid sequence that resulted.
2. Highlight the bases of the nucleotides that were inverted in Figure 8.5 and the change in the amino acid sequence that resulted.
3. Highlight the base of the nucleotide that was deleted in Figure 8.6 and the change in the amino acid sequence that resulted.

***For Practice** continued* ➢

For Practice *continued*

4 Highlight the base of the nucleotide that was inserted in Figure 8.7 and the change in the amino acid sequence that resulted.

5 Copy and complete the sentences below which refer to gene mutation.

A gene mutation occurs when the order of _________ in a DNA strand is _______.

The effect that this has is to change the ________ sequence on the mRNA and this in turn changes the _____________ of the amino acids in the __________. The ___________ of the protein is changed.

6 Which has a greater effect on the structure of the protein that is synthesised: a substitution or a deletion mutation? Give a reason for your answer.

Non-disjunction and effects on karyotype

Figure 8.8 shows **non-disjunction** in a cell with three homologous pairs of chromosomes. Non-disjunction is the failure of a homologous pair of chromosomes to separate during the first meiotic division. Gametes are formed with a chromosome complement that contains one extra chromosome or is one chromosome short.

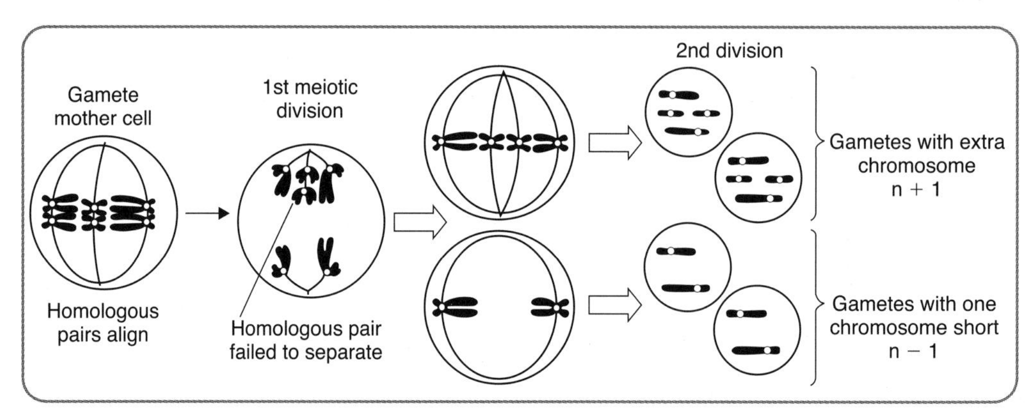

Figure 8.8 Non-disjunction

If a human gamete with an extra chromosome (24) fuses with a normal gamete (23) then a zygote forms with an abnormal chromosome complement of 47. If a human gamete with one chromosome short (22) fuses with a normal gamete (23) then a zygote is formed with an abnormal chromosome complement of 45. Such abnormal chromosome complements give rise to conditions such as Down's, Turner's and Kleinfelter's syndromes.

Figure 8.9 shows the karyotype of a Down's syndrome individual. The karyotype shows that such individuals possess an extra copy of chromosome 21.

Figure 8.10 shows the karyotype of an individual with Turner's syndrome. The karyotype shows that such individuals lack a second sex chromosome.

Figure 8.11 shows the karyotype of a Kleinfelter's individual. The karyotype shows that such individuals possess an extra X-chromosome.

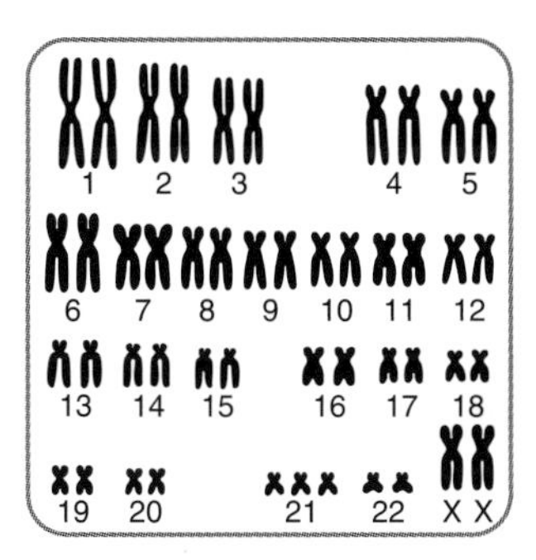

Figure 8.9 Karyotype: Down's syndrome

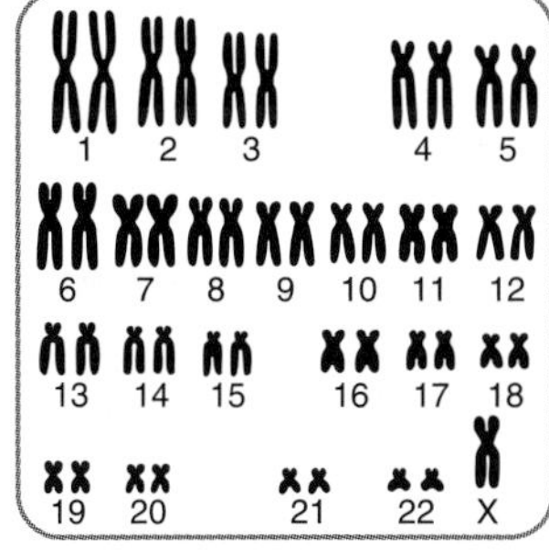

Figure 8.10 Karyotype: Turner's syndrome

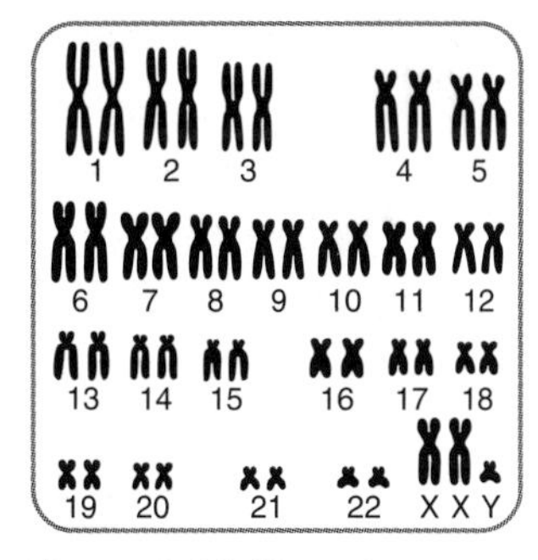

Figure 8.11 Karyotype: Kleinfelter's syndrome

Individuals with Down's syndrome show great variation in degrees of the disorder. In general, people with Down's syndrome have learning difficulties, heart problems and distinctive facial features.

Individuals with Turner's syndrome are female, short in stature and show a lack of sexual development at puberty.

Individuals with Kleinfelter's syndrome are male with testes that remain small. Few sperm are produced and adults are usually infertile. Due to the small testes, low levels of testosterone are produced, so muscle development during puberty is reduced. Treatment to counteract these effects involves supplementing testosterone during puberty.

Hints and Tips

Non-disjunction occurs during the formation of both sperm and ova.

The chance of a non-disjunction increases with the age of the individual.

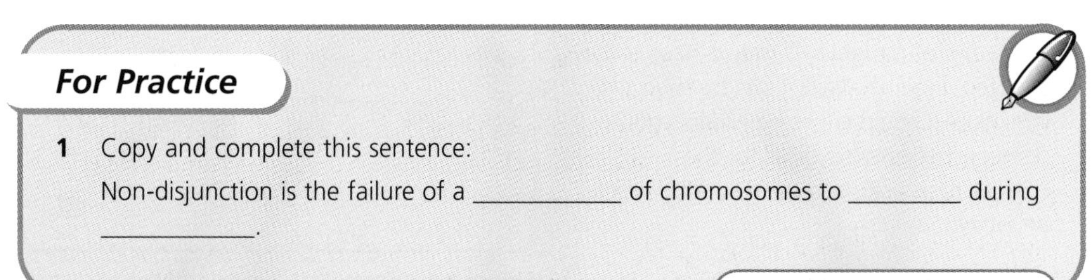

For Practice

1 Copy and complete this sentence:

Non-disjunction is the failure of a ___________ of chromosomes to ________ during ___________.

For Practice *continued* ➢

For Practice continued

2 State the chromosome complement of the gametes formed in humans when a single non-disjunction occurs during meiosis.

3 Explain how an individual inherits a chromosome complement of 47 that contains an extra 21st chromosome.

Genetic screening and counselling

The use of family histories in determining genotype

From **family trees**, genetic counsellors are able to calculate the chance of an individual inheriting a condition or of passing a condition on to their children. Genetic conditions of importance that are determined through family trees include the autosomal recessive conditions of cystic fibrosis, albinism and phenylketonuria; the autosomal dominant condition of Huntington's chorea; and the sex-linked conditions of haemophilia and muscular dystrophy.

Individuals with cystic fibrosis produce thick mucus in the lungs and digestive tract.

Individuals with albinism fail to produce skin and eye pigments.

Individuals with phenylketonuria fail to produce the enzyme that converts the amino acid phenylalanine to tyrosine. If not treated, the build-up of the abnormal metabolic products of phenylalanine leads to brain damage shortly after birth.

Individuals with Huntington's chorea lose muscle and motor co-ordination. The symptoms do not usually appear until middle age.

Individuals with haemophilia bleed excessively as the blood fails to clot as normal.

Individuals with muscular dystrophy show progressive weakness and degeneration in the muscles that control movement.

Figure 8.12 shows a family tree for the incidence of the autosomal recessive condition cystic fibrosis.

Evidence shows that the allele is recessive as the parents of C, F, M and S did not have the disorder. They must be heterozygous. Individuals C and F show that the allele is not sex-linked. If the allele was sex-linked the father of daughter C would have to be affected. From the tree it can be seen that individual R could either be homozygous dominant or heterozygous for the condition, as both parents (N and P) were heterozygous.

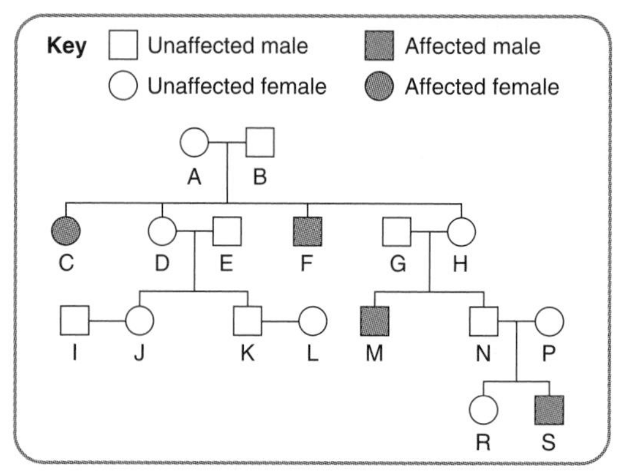

Figure 8.12 Family tree: cystic fibrosis

Figure 8.13 shows a family tree for the incidence of the autosomal dominant condition Huntington's chorea.

Evidence shows that the allele is dominant as each sufferer has a parent with the condition and when a branch of the family (G) does not express the condition then it fails to appear in future generations of that branch. Individuals D and G show that it is not sex-linked as a female with the condition (D) did not pass the condition to her son (G). From the tree it can be seen that individuals O and Q are heterozygous for the condition and that the chances are that 1 in 2 of their children would inherit the disorder.

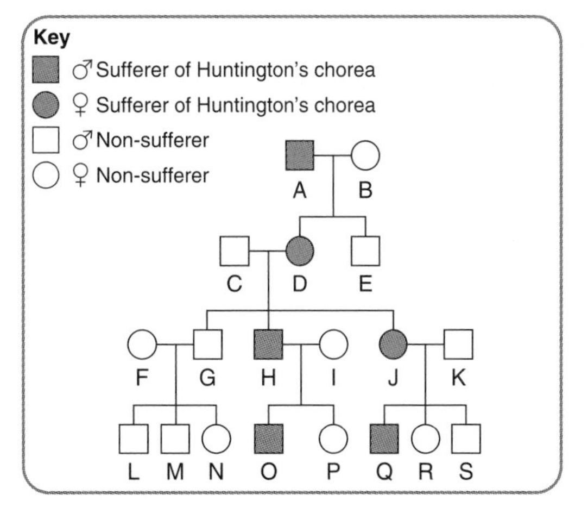

Figure 8.13 Family tree: Huntington's chorea

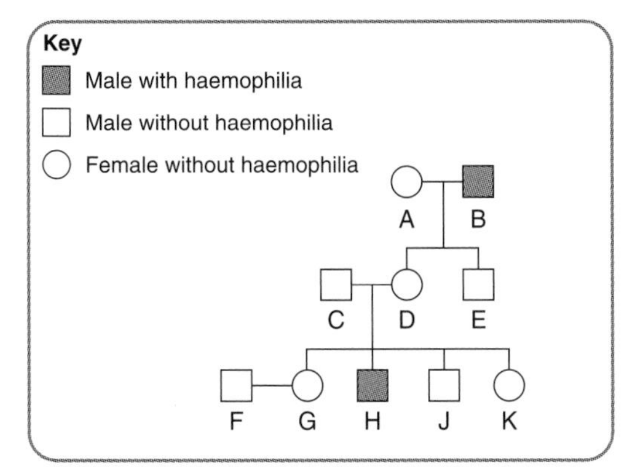

Figure 8.14 Family tree: haemophilia

Figure 8.14 shows a family tree for the incidence of the sex-linked recessive condition haemophilia.

Evidence shows that this is a sex-linked disorder as males only are affected (B and H) and the son (E) of an affected male did not show the condition. The condition misses a generation but is carried by a female (D). From the tree the females G and K could be either homozygous dominant or a carrier for the condition as their mother (D) was a carrier.

Use of karyotypes of fetal material

If there is the possibility of a genetic disorder in the fetus, fetal cells may be obtained by amniocentesis, in which amniotic fluid is extracted, or by chorionic villus sampling, in which a sample of the placenta is taken. The cells are photographed and the chromosome complement of the fetus is shown in the karyotype produced.

The karyotype provides evidence of any chromosomal abnormalities (see Figures 8.9, 8.10 and 8.11).

Risk evaluation in cases of polygenic inheritance

The risks associated with polygenic inherited disorders are difficult to evaluate with any degree of accuracy. Data gathered on an individual is matched against data that is based on previous case histories. A final risk assessment is produced from statistical calculations based on comparisons between the data on the individual and the historic data.

Polygenic inherited disorders include conditions such as cleft palate, congenital heart disease and schizophrenia.

Post-natal screening for conditions which have a genetic basis

Shortly after birth babies are tested for the genetically inherited disorder phenylketonuria (PKU). If diagnosed, the individual is put onto a diet that contains the minimal phenylalanine required for normal growth and development, together with a tyrosine supplement. Figure 8.15 outlines detail on the PKU condition.

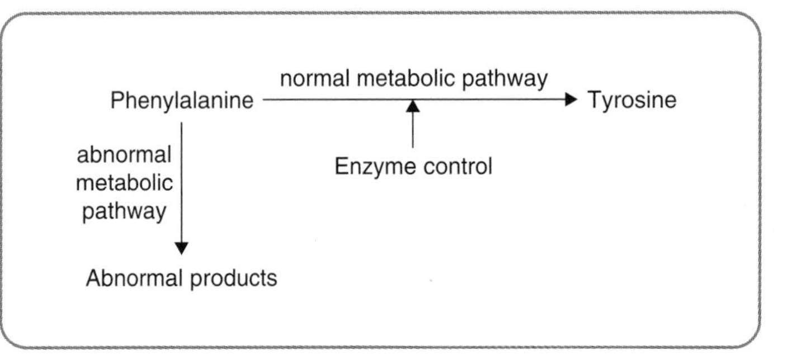

Figure 8.15 PKU condition

The normal metabolic pathway is under the control of an enzyme. The amino acid phenylalanine fits to the active site of the enzyme and is converted to tyrosine. As a result of a gene mutation, the shape of the enzyme produced is altered and phenylalanine no longer fits to the active site. An individual homozygous for this condition no longer produces an enzyme to convert phenylalanine to tyrosine. The abnormal metabolic pathway is now followed and the abnormal products of this metabolism are toxic to developing brain tissue.

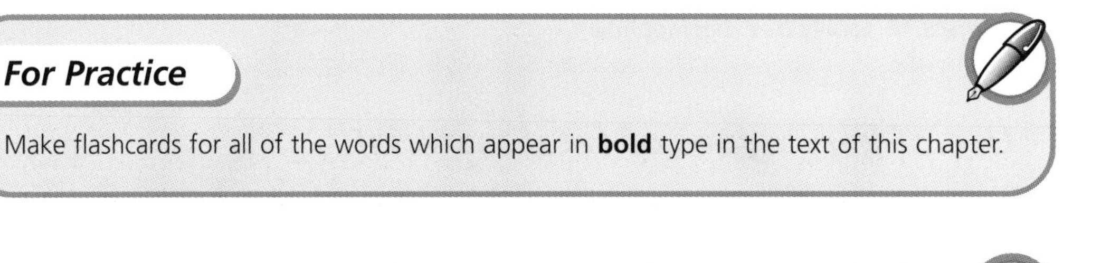

For Practice

Make flashcards for all of the words which appear in **bold** type in the text of this chapter.

Exam Questions

1 Red–green colour deficiency is a sex-linked condition caused by a recessive allele (X^b). The genotypes of a family are shown:

Parents $X^B X^b$ x $X^b Y$

↓

Offspring (Jim) $X^b Y$

Which of the following shows the genotypes of Jim's grandparents?

Exam Questions continued

	Maternal grandmother	Maternal grandfather	Paternal grandmother	Paternal grandfather
A	$X^B X^b$	$X^B Y$	$X^B X^b$	$X^B Y$
B	$X^B X^B$	$X^B Y$	$X^B X^b$	$X^B Y$
C	$X^B X^b$	$X^b Y$	$X^B X^B$	$X^b Y$
D	$X^B X^B$	$X^b Y$	$X^B X^B$	$X^b Y$

2 A genetic disorder results in males having the sex chromosomes XXY. The disorder that led to this condition was caused by:

A chiasmata in the sex chromosomes leading to crossing-over
B a disorder in a sex-linked gene
C non-disjunction during meiosis
D an ovum being fertilised by two sperm cells.

3 The gene for blood groups A, B, AB and O has three alleles. Alleles A and B are co-dominant to allele O.

Which line in the table below identifies a family in which one of the children is related to only one of the parents?

	Blood groups of parents	Blood groups of children
A	AB and O	A and B
B	AB and B	B and O
C	A and B	B and O
D	AB and B	A and A

4 In humans, the bone disorder achondroplasia is an inherited condition in which the arms and legs remain short and the torso is of normal length. The allele for the disorder (A) is dominant to the allele for normal growth (a).

Figure 8.16 shows the family tree of the inheritance of achondroplasia over three generations.

(a) What evidence supports the statement that the condition is due to a dominant allele? *(2)*

(b) Identify the genotype of the affected female F and justify your answer. *(2)*

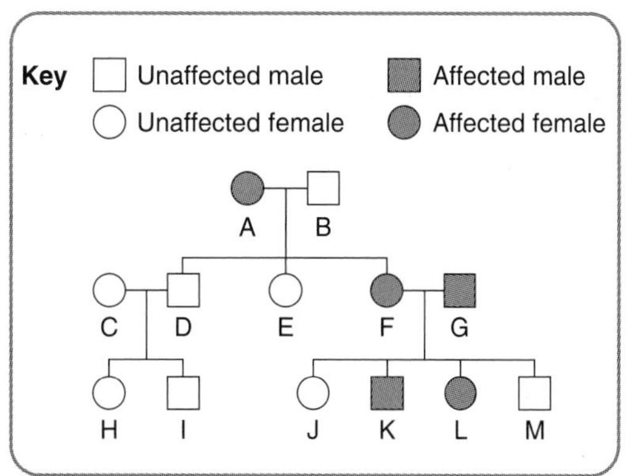

Figure 8.16 Family tree: achondroplasia

Exam Questions continued

(c) Individuals F and G have another child. What is the chance that the child will inherit the condition? *(1)*

5 Insulin is made up of two protein chains. The protein chains are held together by cysteine to cysteine cross bridges.

Figure 8.17 shows the amino acids in part of a normal insulin molecule.

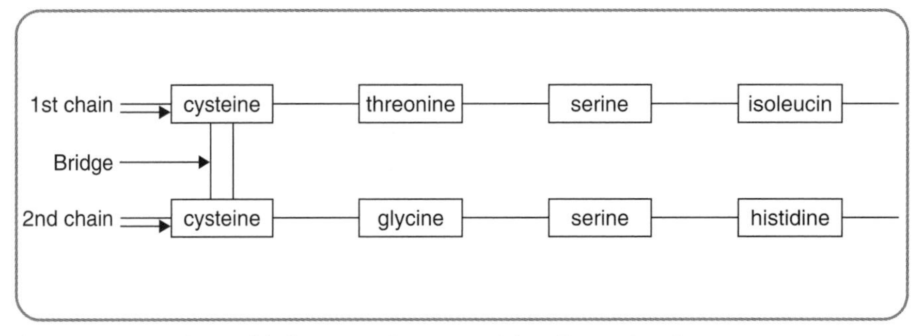

Figure 8.17 Amino acids in part of a normal insulin molecule

(a) Name the bond that holds the amino acids together in the protein chain. *(1)*

(b) The table below shows the mRNA codons for the amino acids in the diagram.

Amino acid	Codon
cysteine	UGU
glycine	GGG
histidine	CAU
isoleucine	AUC
serine	AGU
threonine	ACA

Write the sequence of bases of the tRNA anticodons that correspond to the sequence of amino acids shown in the 1st protein chain of insulin. *(1)*

(c) If a mutation occurred that replaced the first uracil base in the codon for cysteine with the base adenine, describe **two** ways in which the structure of the insulin molecule would be changed. *(2)*

Answers

1 Tracing back from Jim's parents:

Choice B for the maternal grandmother does not possess the recessive allele. On the paternal side the recessive allele must come from the grandmother – this is not shown by C or D. **Answer** A

2 Male has inherited an extra sex chromosome. One of gametes must have had an extra chromosome. This is caused by non-disjunction in meiosis. **Answer** C

3 All possible genotypes of parents must be used and all crosses carried out.

Possible genotypes of parents	Possible genotypes of children
Line A AB and OO	AO or BO **OK**
Line B AB and BB or BO	AB or BB or AO or BO **NOT OO**
Line C AA or AO and BB or BO	AB or AO or BO or OO **OK**
Line D AB and BB or BO	AB or BB or AO or BO **OK**

Answer B

4 (a) It is easier to show that it is not recessive rather than to show it is dominant and if it is not recessive then it must be dominant. If recessive, **all** offspring of F and G would be affected (1 mark).

F and G have unaffected offspring, therefore it must be dominant (1 mark).

(b) Aa

She received the dominant allele from her affected mother (A) (1 mark) and the recessive allele (a) from her homozygous recessive father (B) (1 mark).

(c) Carry out the cross

P genotype	Aa	X	Aa
Gametes	A or a		A or a
Possible F1 Genotypes	AA	2Aa	aa
Possible F1 Phenotypes	3 achondroplasia : 1 normal		

Four possible outcomes of which three inherit the condition. **Answer** 3 in 4 or 75%.

5 (a) You should know that amino acids are held together by **peptide bonds**.

(b) Do this in the correct sequence: amino acid sequence then mRNA codon sequence then tRNA anticodon sequence.

Amino acids	cysteine	threonine	serine	isoleucine
mRNA codons	UGU	ACA	AGU	AUC
tRNA anticodons	**ACA**	**UGU**	**UCA**	**UAG**

(c) Original codon = UGU, after mutation = AGU. This is the codon for the amino acid serine.

1 Serine would replace cysteine.

2 With a serine in place for the cysteine the bridge to the other cysteine does not form.

Chapter 9

REPRODUCTION AND DEVELOPMENT

Structure and function of reproductive organs

Females

The female reproductive system consists of the ovaries, oviducts, uterus, cervix and vagina.

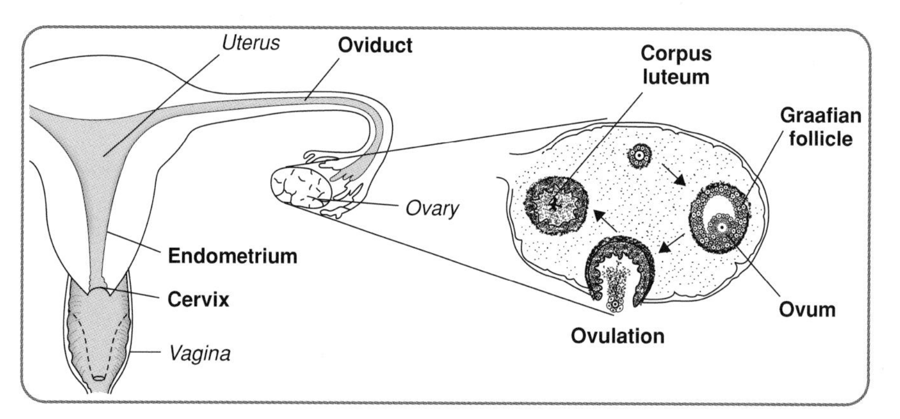

Figure 9.1 Female reproductive system

Ova, the female gametes, are produced in the **ovaries**. Each month an ovum matures in a Graafian follicle and is released at ovulation. As the ovum is moved along the **oviduct** it may meet a sperm and be fertilised. If so, the resulting ball of cells embeds itself in the lining of the uterus called the **endometrium** and develops into a fetus. If not, it is lost with the endometrium during menstruation. After ovulation the Graafian follicle changes its function and develops in the **corpus luteum**.

For Practice

Identify these stages in follicle development:

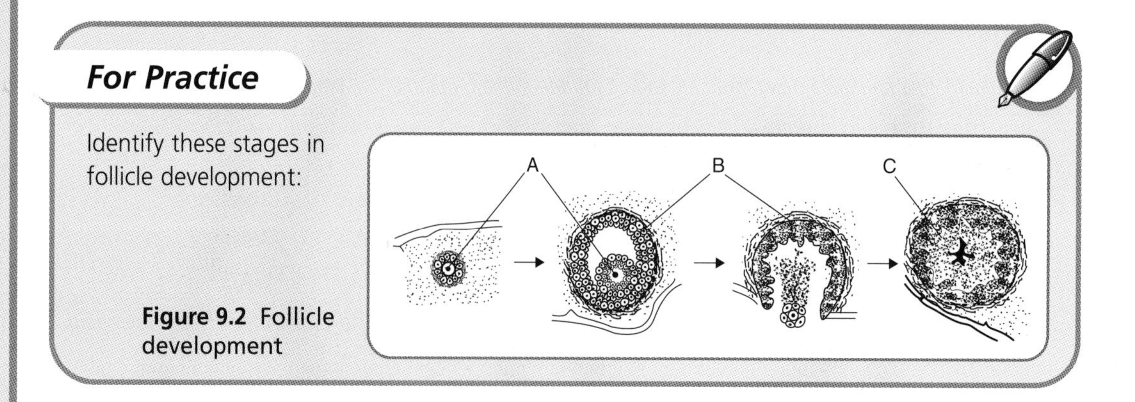

Figure 9.2 Follicle development

Males

The male reproductive system consists of the testes, sperm ducts, penis and the associated glands known as the prostate gland and the seminal vesicles.

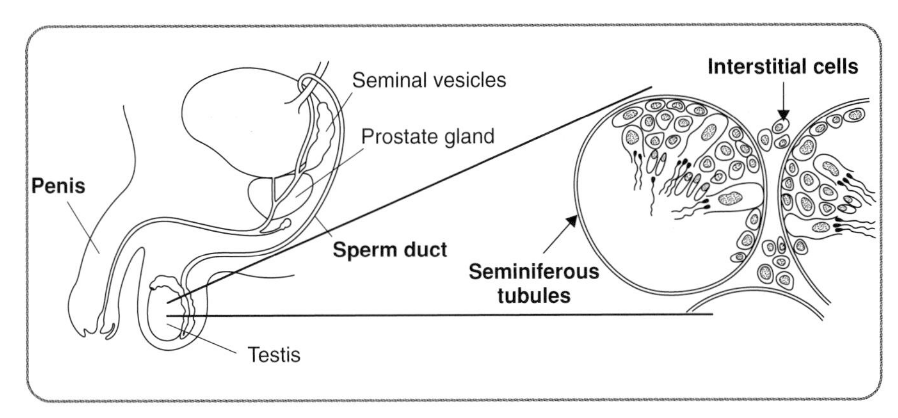

Figure 9.3 Male reproductive system

Sperm, the male gametes, develop in the **seminiferous tubules** in the testes. The hormone **testosterone** is produced in the interstitial cells between the seminiferous tubules.

The **prostrate gland** and **seminal vesicles** produce fluids that help sperm swim towards the ova because they:

- neutralise the acid conditions in the vagina because they are alkaline
- provide fructose to give the sperm energy for swimming
- contain hormones which stimulate contractions in the oviduct walls which push the sperm along.

Hormonal control

The **pituitary gland** produces two 'sex' hormones called **luteinising hormone** (LH) and **follicle stimulating hormone** (FSH) that affect reproduction in both males and females.

Males

LH stimulates the interstitial cells in the testes to produce the hormone testosterone. FSH and testosterone combine to stimulate the production of sperm. LH secretion and thus sperm production is continuous throughout adult life.

Females

Pituitary FSH and LH and **oestrogen** and **progesterone** from the ovary have a 'conversation' to control the menstrual cycle. Figure 9.4 shows the sequence of the interaction.

1 Pituitary FSH stimulates an ovum to mature inside a developing Graafian follicle.

2 The Graafian follicle secretes oestrogen which:
   - inhibits the production of FSH by the pituitary
   - stimulates the uterus to start growing an endometrium
   - stimulates the cervix to secrete thin mucus ideal for sperm to swim in
   - stimulates the pituitary to produce LH.

3 A surge of LH from the pituitary causes:
   - Ovulation. The ovum is released from the ovary into the oviduct
   - the Graafian follicle to develop into the corpus luteum.

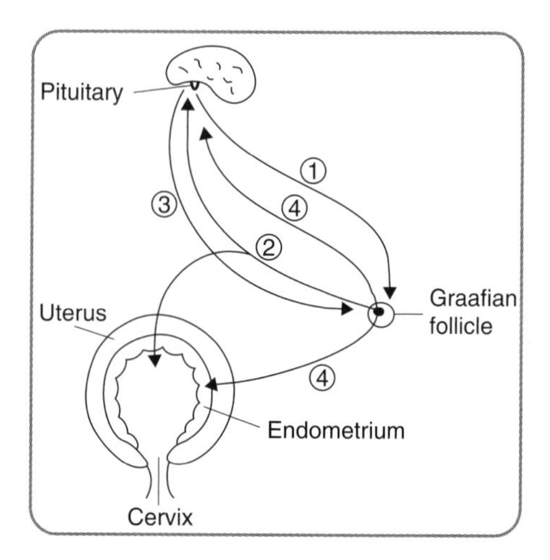

Figure 9.4 Hormone control

4 The corpus luteum produces progesterone and oestrogen which cause:
   - the uterus lining, known as the endometrium, to thicken ready for a fertilised ovum
   - the cervix mucus to become thick and sticky to prevent the entry of bacteria
   - inhibition of the production of FSH and LH by the pituitary.

What happens next? The table below shows the two possible options.

If the ovum is fertilised …	If the ovum is not fertilised …
the corpus luteum continues to produce progesterone and oestrogen for several months until the placenta is sufficiently developed to take over full production	the corpus luteum degenerates after about 12 days
progesterone and oestrogen from the corpus luteum and placenta retain the endometrium	when progesterone levels fall the endometrium dies and menstruation starts
progesterone and oestrogen from the corpus luteum and placenta inhibit FSH until birth	low levels of progesterone allow FSH production and the menstrual cycle begins again

Hints and Tips

Sex hormones with long names ending with 'hormone' are produced in the pituitary gland. Because they have long names these are usually reduced to letters – FSH and LH.

Progesterone is a favourite hormone in the exam. Remember:

- progesterone is produced by the corpus luteum *and* the placenta
- progesterone inhibits FSH and retains the endometrium (uterus lining)
- progesterone is in contraceptive pills as it inhibits FSH production.

For Practice

For Practice

Complete a table like the one below with ticks (✓) to help you memorise the function of the reproductive hormones for both males and females.

Hormone function	Hormones			
	FSH	Oestrogen	LH	Progesterone
Produced by the Graafian follicle to inhibit FSH production		✓		✓
Produced by the corpus luteum to stimulate endometrium thickening		✓		✓
Causes cervical mucus to thicken		✓		✓
Stimulates LH production				
Promotes ovulation				
Stimulates interstitial cells to produce testosterone				
Produced by the placenta to retain the endometrium				
Both inhibit FSH production		✓		
Works with testosterone to stimulate sperm production	✓			
Causes cervical mucus to thin	✓			
Stimulates an ovum to grow and mature inside a Graafian follicle		✓		

Intervention in fertility

In males, infertility may be due to a low sperm count or sperm that are inactive or abnormal in some way. Treatments include supplements of testosterone to boost the sperm numbers and activity, or artificial insemination using healthy sperm donated by another male.

Female infertility may be due to ovulation failure which is treated with either FSH or LH. A blockage of the oviduct is treated by surgery or by ***in vitro* fertilisation**. During *in vitro* fertilisation sperm and ova are mixed in a 'test-tube' outside the female's body and then one of the developing embryos is implanted in the uterus. Fertility drugs which stimulate the growth of the endometrium are used to treat implantation failure.

For Practice

Complete these sentences:

Female infertility may be due to blocked oviducts which is a problem that may be overcome by ____ ________ fertilisation. Male infertility can be due to a reduced number of sperm which can be treated by ________________ hormone injections.

Conception is most likely to happen during a fertile period which is approximately five days around the time of ovulation. As can be seen from Figure 9.5, this is normally from three days before ovulation to one day after. This is because sperm live for about three days and an ovum is viable for about one day. The time of ovulation can be roughly estimated to be 16 days after menstruation but this is variable. Couples who do not wish to conceive should have intercourse during the infertile or 'safe period'.

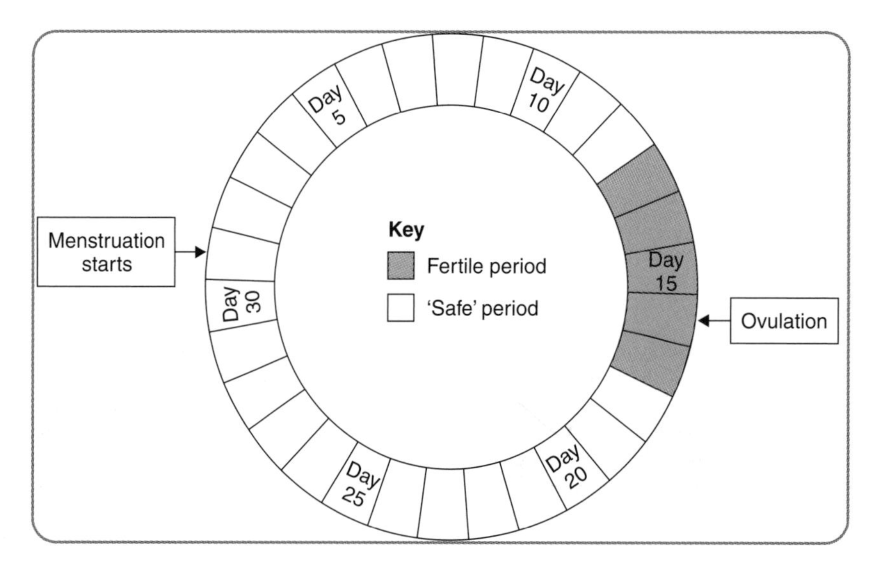

Figure 9.5 Fertile period

As can be seen from Figure 9.6, a slight rise in waking body temperature on the day after ovulation and thinning of the cervical mucus are more reliable indications of the time of ovulation.

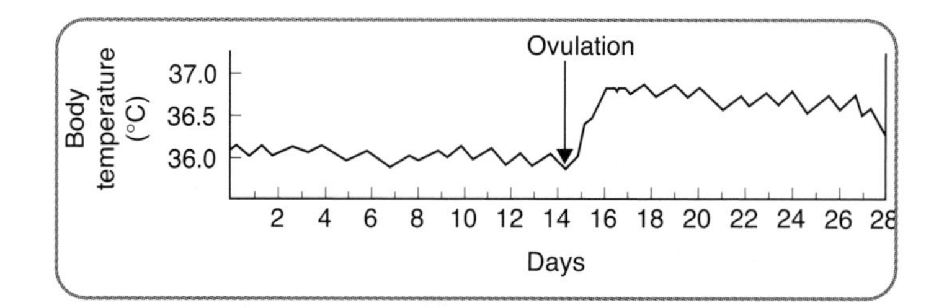

Figure 9.6 Ovulation

For Practice

1 (a) From Figure 9.6, on which day after menstruation does ovulation occur.
(b) What is the approximate rise in temperature at ovulation?

2 Complete these sentences:
The fertile period is approximately five days around ____________. The best indication is a rise in ________________ and the thinning of the ______________ mucus.

Contraceptive methods

These depend on preventing the sperm reaching the uterine tubes or preventing ovulation. Condoms, intercourse during the 'safe period', and sterilisation all prevent sperm reaching the ova. The **contraceptive pill** prevents ovulation by 'pretending' pregnancy with a mixture of progesterone and oestrogen which inhibit FSH production by the pituitary. Without FSH an ovum cannot grow and mature in the ovary and pregnancy is impossible.

For Practice

Copy and complete the table below with ticks (✓) to show if the contraceptive method prevents sperm reaching the ova, or prevents ovulation. It is likely that learning one example of each will be sufficient – so try to memorise the example of each that you are most likely to remember.

Contraceptive method	Prevents sperm reaching ova	Prevents ovulation
Condom		
Vasectomy (cutting sperm duct)		
Laparotomy (cutting oviduct)		
Contraceptive pill		
Sexual intercourse during the 'safe period'		
Spermicidal gel		

Development

Intra-uterine development

The first cell produced from the fertilisation of an ovum by a sperm in the oviduct (uterine tube) is called a **zygote**. The zygote divides repeatedly to produce a ball of cells by a process known as **cleavage**. **Dizygotic twins** (not identical) result from two ova being fertilised by different sperm. **Monozygotic** (identical) twins result from the group of cells from one

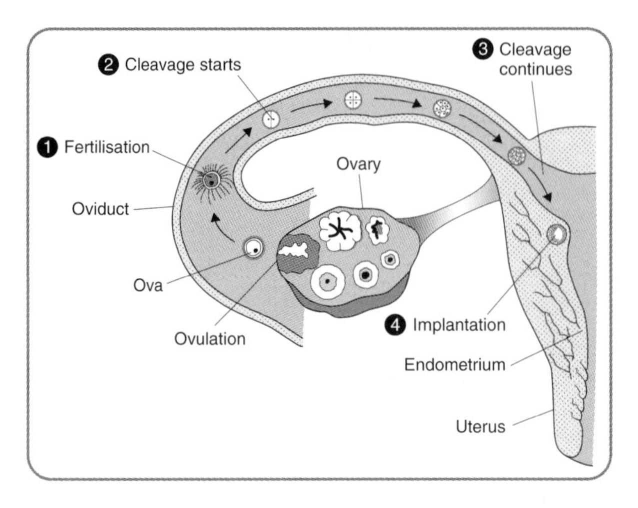

Figure 9.7 Intra-uterine development

fertilised ovum splitting into two during cleavage. Once the ball of cells reaches the uterus it embeds itself into the endometrium – this is **implantation**. Nourished by the endometrium the embryonic cells continue dividing and also differentiate to form the specialised cells which form the various organs of the fetus.

The exchange of materials between the blood of the fetus and that of the mother occurs across the placenta. Figure 9.8 shows the relationship between the mother's circulatory system and that of the developing embryo.

Oxygen diffuses from the maternal blood into the fetal blood helped by the fact that the fetal haemoglobin has a greater affinity for oxygen. Carbon dioxide diffuses from the fetal blood to the maternal blood down a concentration gradient. Fetal blood obtains glucose by **active transport** and antibodies by **pinocytosis** which give passive immunity. Harmful substances can also cross the placenta to the fetus, for example, nicotine, alcohol, heroin and pathogens such as rubella and HIV viruses.

The placenta takes over the production of progesterone and oestrogen from the corpus luteum. Progesterone and oestrogen maintain the endometrium and inhibit ovulation for the duration of pregnancy. Under their influence the breasts grow and prepare for milk production.

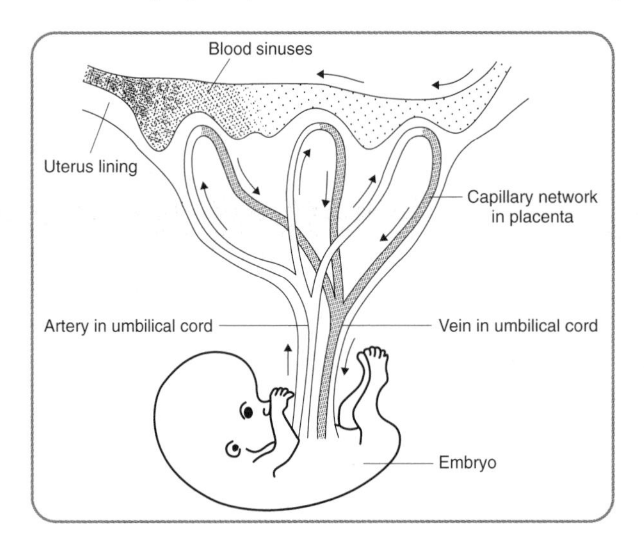

Figure 9.8 Embryo development

Normally the mother's immune system does not attack the fetus despite its different antigen signature. If the mother is Rhesus negative and her baby is **Rhesus positive** there is a chance of some blood mixing at birth. The immune system of the mother then makes anti-D antibodies and memory cells that wait for a future 'invader'. A second Rhesus positive fetus is attacked by anti-D antibodies through the placenta. The fetus can be saved by replacing the Rhesus positive blood with Rhesus negative blood. Nowadays, the mother is injected with anti-D antibodies, just after the first birth, to destroy the D antigens and prevent the immune response.

For Practice

The allele for Rhesus positive is dominant (D) to the Rhesus negative allele (d). A Rhesus negative mother and a heterozygous Rhesus positive father already have one Rhesus positive child and are expecting a second child. What is the probability that the second child will be Rhesus positive and set off an immune reaction?

Birth

The concentration of oestrogen produced by the placenta rises throughout pregnancy. When it reaches a critical concentration it stimulates the pituitary gland to produce the hormone **oxytocin** which stimulates the muscular walls of the uterus to contract rhythmically. Early in labour the amniotic sac ruptures, releasing the 'waters', and the cervix dilates. The rhythmic contractions become more frequent and powerful until the baby is forced out. Soon after birth the placenta or 'afterbirth' is also expelled. Artificial oxytocin is used to induce the birth of a baby who is regarded as being well past full term.

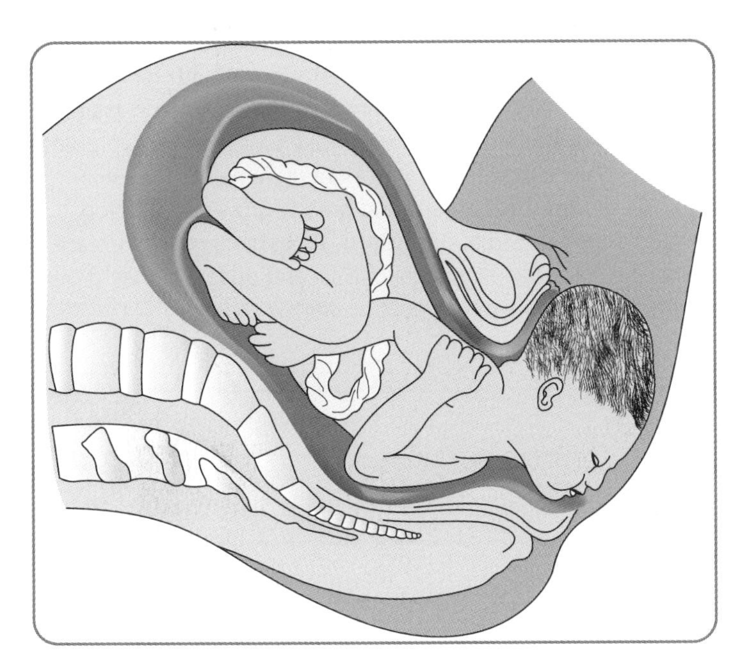

Figure 9.9 Birth

Without the placenta the inhibitory effect of progesterone is lost. The pituitary produces **prolactin** which stimulates milk production. In the first few days after birth the breasts produce **colostrum**. Colostrum contains many antibodies to provide passive immunity and also contains a higher concentration of protein, minerals and vitamins than breast milk. Breast milk also contains antibodies and so many regard it as superior to dried modified cow's milk. However, breast milk may contain harmful fat soluble chemicals eaten by the mother in food, e.g. pesticides.

The pattern of growth after birth

Growth after birth is rapid and the rate declines slowly until it stops in the early twenties. There is a 'growth spurt' at **puberty** which, on average, happens earlier in girls. Different parts of the body mature at different ages – there is a change in body proportions with age. At birth the head is relatively large and reduces in proportion as the torso and limbs grow. The reproductive organs are the last to mature.

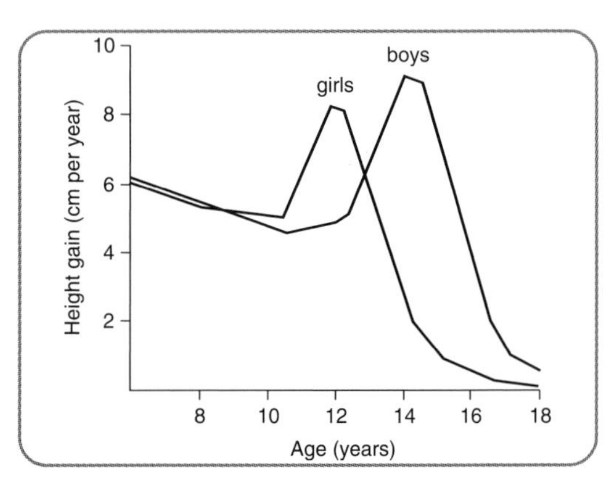

Figure 9.10 Human development

For Practice

What is the average age of the growth spurt in males and females?

Growth hormone (GH) produced by the pituitary gland promotes the entry of amino acids into cells, protein synthesis and cell division and so accelerates the growth of bones and muscles. At puberty FSH and LH begin to be produced by the pituitary in both females and males. These stimulate the interstitial cells in the testes to produce testosterone in males, and the ovaries in females to produce oestrogen and progesterone which, in turn, promote the development of the secondary sexual characteristics. At puberty females increase in height and weight, hips widen, the reproductive organs and breasts grow and develop, pubic and underarm hair grows and the menstrual cycle starts. At puberty males increase in height, weight and musculature, the voice deepens, pubic, facial and underarm hair grows, reproductive organs grow and develop and sperm production starts.

For Practice

Make flashcards for all of the words that appear in **bold** type in the text of this chapter.

Exam Questions

1 Figure 9.11 shows a 30-day menstrual cycle.

Ovulation takes place on day

A 1

B 14

C 16

D 30

Figure 9.11 30-day menstrual cycle

2 Which of the following is a chemical substance that passes by active transport from maternal to fetal blood in the placenta?

A Oxygen

B Glucose

C Carbon dioxide

D Antibodies

3 The sequence of events in pre-natal development is:

A cleavage → implantation → differentiation

B cleavage → differentiation → implantation

C implantation → differentiation → cleavage

D implantation → cleavage → differentiation.

4 The Rhesus factor increases the risk of a fetus being rejected by the maternal immune system. Which line in the table below identifies the conditions that give the greatest risk of rejection?
RH^+ = Rhesus factor present; RH^- = Rhesus factor absent

	Fetal blood type	Maternal blood type	Number of pregnancies
A	Rh^+	Rh^-	First
B	Rh^+	Rh^-	Second
C	Rh^-	Rh^+	First
D	Rh^-	Rh^+	Second

Exam Questions *continued*

5 The role of the hormone oxytocin is to:

A stimulate ovulation

B maintain the endometrium

C stimulate the production of milk

D initiate contraction of the uterus.

6 Figure 9.12 shows changes in the ovary during a menstrual cycle.

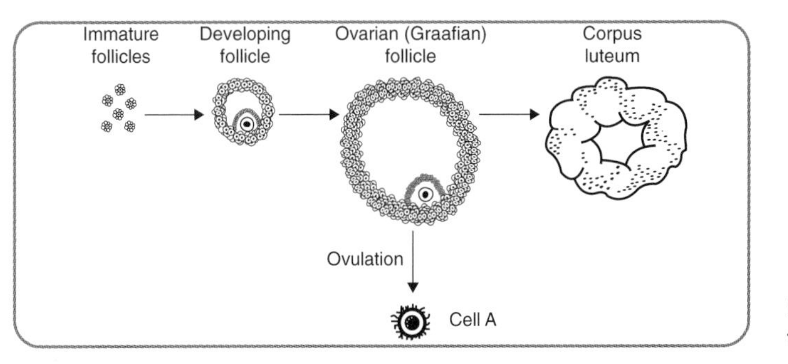

Figure 9.12 Changes in the ovary

(a) Copy and complete this table which relates to hormones in the cycle. *(3)*

Function	Hormone
To stimulate development of a Graafian follicle	
To bring about ovulation	
To stimulate the corpus luteum to produce progesterone	

(b) The corpus luteum produces progesterone during the first weeks of pregnancy. Where is progesterone produced during the later stages of pregnancy? *(1)*

(c) State two effects of a lowering of progesterone levels at the end of a menstrual cycle. *(2)*

7 Give an account of infertility under the following headings:

(a) Causes of infertility *(4)*

(b) Treatment of infertility *(3)*

(c) Methods of calculating the fertile period *(3)*

Answers

1 Luteinising hormone stimulates ovulation and the diagram shows that peak LH production is on day 16. **Answer** C

2 This is one of those questions where you either know the fact or you have to whittle away the ones you know are wrong until you reach one that must be the right answer. Carbon dioxide goes from fetal to maternal blood by diffusion. Oxygen diffuses and antibodies cross by pinocytosis. Glucose passes from the maternal to the fetal blood supply by active transport just as it does in the small intestine. **Answer** B

3 Cleavage is the repeated division of the fertilised ovum, or zygote, until a ball of cells is produced. Implantation is when the ball of cells embeds itself into the uterus lining where it gains nourishment for further growth and development which includes cells becoming different to do different jobs. **Answer** A

4 A Rhesus positive (Rh^+) fetus has antigens that are recognised by the immune system of a Rhesus negative (Rh^-) mother as being non-self. However, the immune system of the mother only becomes activated if blood mixes at the first birth. The rejection reaction occurs during the second Rh^+ pregnancy. **Answer** B

5 Again, this is one of those questions where you either know the fact or you have to whittle away the ones you know are wrong until you reach one that must be the right answer. Ovulation is stimulated by LH, progesterone maintains the endometrium and prolactin stimulates the production of milk. Oxytocin induces birth by stimulating the muscular walls of the uterus to contract. **Answer** D

6 (a) To stimulate development of a Graafian follicle = FSH (1 mark)

To bring about ovulation = LH (1 mark)

To stimulate the corpus luteum to produce progesterone = LH (1 mark)

(b) As the placenta develops its produces more and more of the progesterone until it is the sole source and the corpus luteum disintegrates.

(c) If the ovum is not fertilised the corpus luteum ceases progesterone production after about 14 days. The endometrium is shed (menstruation) (1 mark) and FSH production by the pituitary starts (1 mark).

Answers continued

7

Nos	Mark scoring points		Comments
	Causes of infertility		*Write the sub-headings*
1	Female: Failure to ovulate *or* Failure of fertilised egg cell to implant *or* Oviducts blocked *or* Uterine fibroids formed *or* Hormone imbalance	2	*Any two of these points for **1 mark each***
2	Male: Low sperm count *or* Abnormal sperm *or* Impotence *or* Hormone imbalance	2	*Any two of these points for **1 mark each***
			Maximum of 4 marks
	Treatment of infertility		*Write the sub-heading*
3	Artificial insemination	1	
4	Sperm introduced into uterus by method other than sexual intercourse	1	
5	*In vitro* fertilisation	1	*A full description will do instead of the term, such as 'fertilisation takes place outside the body and fertilised egg is inserted'*
6	Use of fertility drugs for male or female	1	*Or appropriate examples such as testosterone for males and FSH or LH in females*
			Maximum of 3 marks
	Methods of calculating the fertile period		*Write the sub-heading*
7	At ovulation the body temperature rises slightly	1	
8	Daily temperature recorded for several menstrual cycles to find the day when ovulation is most likely to occur	1	
9	Fertile period lasts for 3 to 4 days around the rise in temperature	1	
10	Conception unlikely before or after fertile period	1	
11	Cervical mucus thins during the fertile period	1	
			Maximum of 3 marks
			Maximum = 10 marks

Chapter 10

TRANSPORT MECHANISMS

The need for a transport system with vessels

As an organism becomes larger its **surface area to volume ratio** decreases. This means that each cell in a large organism has less skin to itself than a cell in a small organism. Only very small organisms, comprised of just a few cells, can rely on diffusion to absorb and transport essential molecules. Large organisms require a transport system, using vessels, to circulate important molecules at the speed required to ensure an adequate supply to all the cells.

Tissue fluid and lymph

Blood is carried to the tissues in thick-walled arteries which, once they enter an organ, divide into many arterioles which again divide into narrow vessels, with walls one cell thick, called **capillaries**. Blood pressure forces the fluid part of the blood, with small soluble molecules, out of the capillaries into the tissue fluid, leaving behind the blood cells and large plasma protein molecules. The cells exchange molecules with the tissue fluid by diffusion down concentration gradients. Useful molecules such as food and oxygen diffuse into the cells whilst carbon dioxide and wastes diffuse out of them. At the venule end of the capillary bed, some tissue fluid re-enters the capillaries. Capillaries merge together to form venules which again merge to form thin-walled vessels with valves, called **veins**, which carry the blood back to the heart. Excess tissue fluid collects in lymph vessels which have thin walls and valves. Lymph is returned to the heart in the **lymphatic system** by contracting skeletal muscles squeezing it past the next valve.

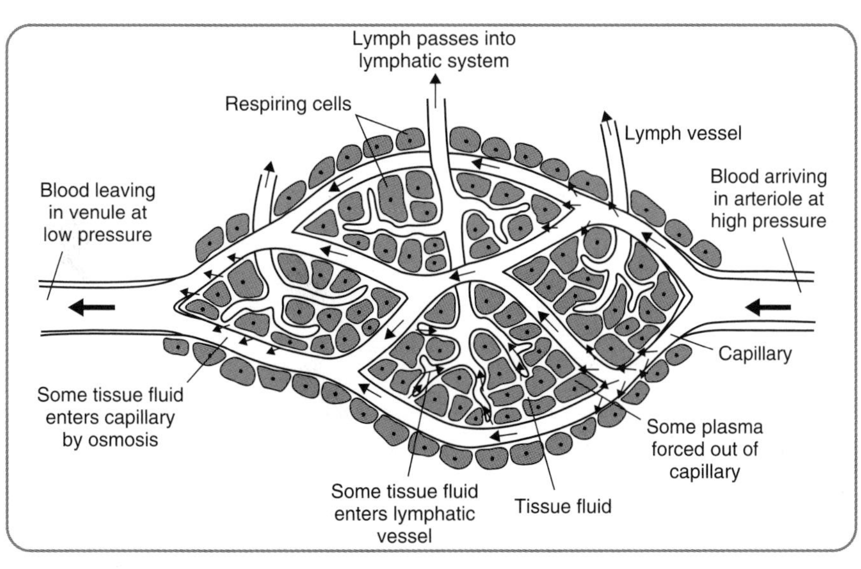

Figure 10.1 Tissue fluid and lymph

For Practice

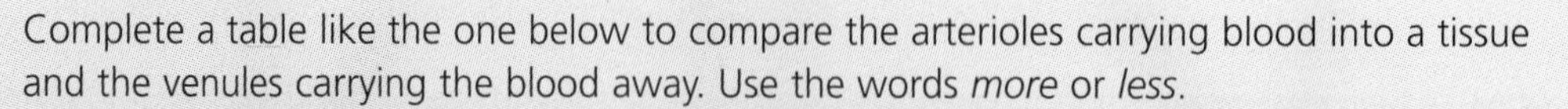

Complete a table like the one below to compare the arterioles carrying blood into a tissue and the venules carrying the blood away. Use the words *more* or *less*.

Feature	Arteriole	Venule
Pressure		
Glucose		
Oxygen		
Carbon dioxide		
Waste		

The need to circulate fluid in vessels

The heart has four chambers, with the right side collecting blood from the body and pumping it to the lungs; whilst the left side collects blood from the lungs and pumps it to the body. The walls of the heart are made of a unique type of muscle called **cardiac muscle** which can contract rapidly, without fatigue, for a lifetime. Deoxygenated blood returning from the body, via the **vena cava**, fills the right atrium until sufficient pressure forces open the **atrioventricular valve** (**tricuspid**) and blood flows into the right ventricle; at this point the atrium contracts, forcing all the blood in. Once full, the right ventricle's muscular walls contract, closing the tricuspid and forcing the blood up through the semilunar valves and on through the **pulmonary artery** to the lungs.

Oxygenated blood returning from the lungs, via the **pulmonary vein**, fills the left atrium until sufficient pressure forces open the **atrioventricular valve** (**bicuspid**) and blood flows into the left ventricle; at this point the atrium contracts, forcing all the blood in. When full, the left ventricle's muscular walls contract, closing the bicuspid valve, and force the blood up through the **semilunar valves** and on through the **aorta** to all the organs of the body.

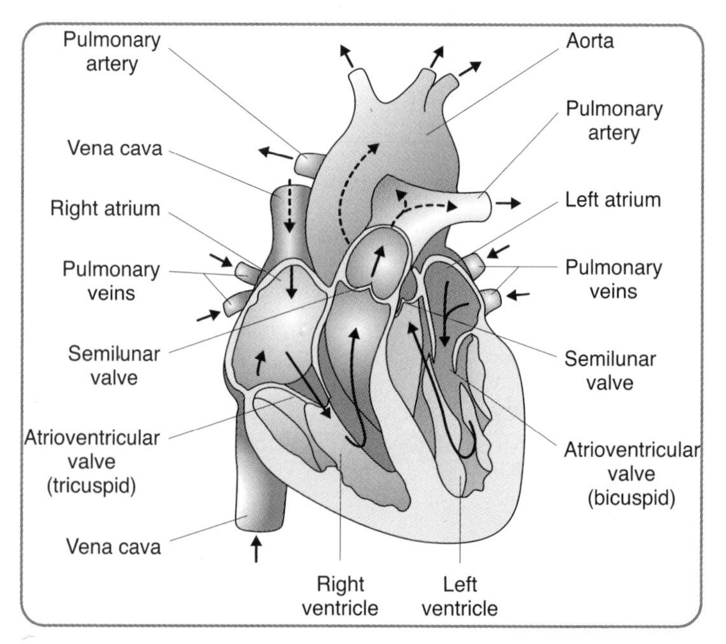

Figure 10.2 Heart

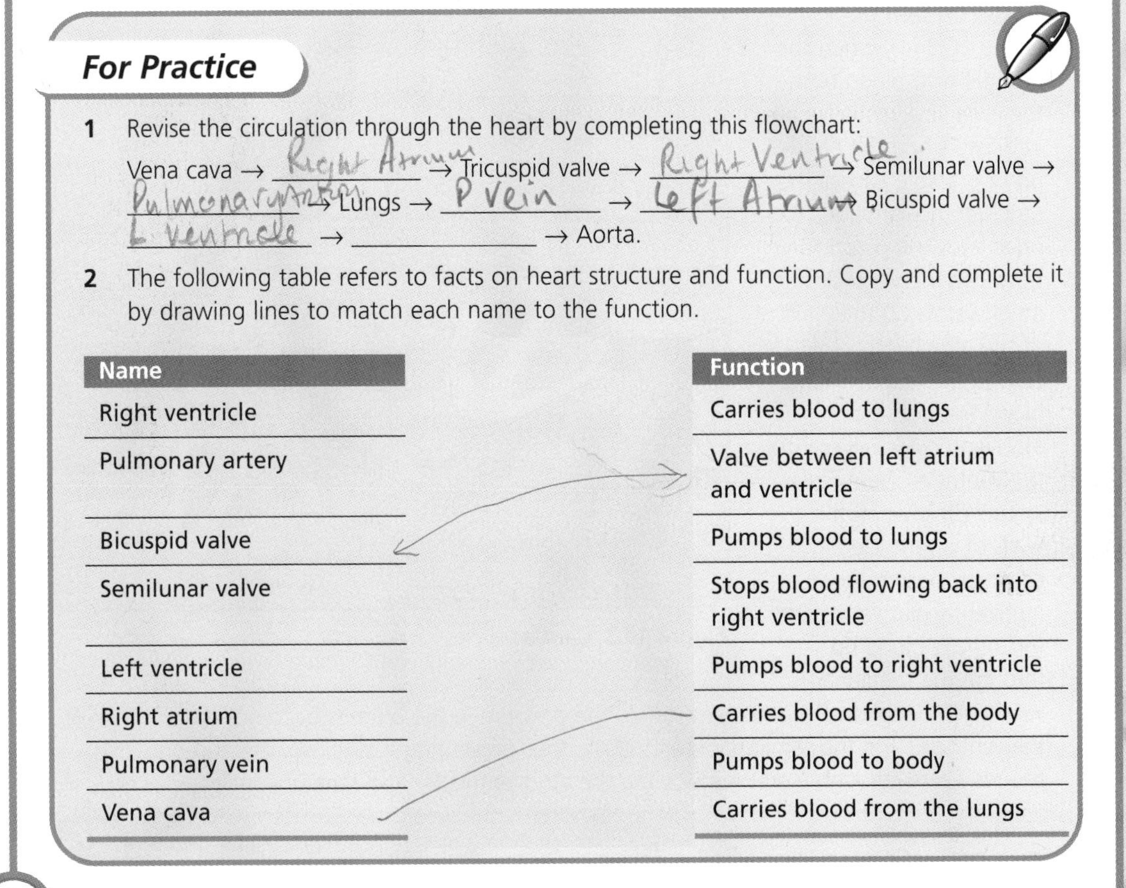

For Practice

1 Revise the circulation through the heart by completing this flowchart:
Vena cava → ____________ → Tricuspid valve → ____________ → Semilunar valve → ____________ → Lungs → ____________ → ____________ → Bicuspid valve → ____________ → ____________ → Aorta.

2 The following table refers to facts on heart structure and function. Copy and complete it by drawing lines to match each name to the function.

Name	Function
Right ventricle	Carries blood to lungs
Pulmonary artery	Valve between left atrium and ventricle
Bicuspid valve	Pumps blood to lungs
Semilunar valve	Stops blood flowing back into right ventricle
Left ventricle	Pumps blood to right ventricle
Right atrium	Carries blood from the body
Pulmonary vein	Pumps blood to body
Vena cava	Carries blood from the lungs

Blood vessels and their functions

The heart muscle has its own blood supply brought in by the **coronary artery**. The **carotid artery** supplies blood to the head and blood returns to the heart in the **jugular vein**. The kidney receives blood from the **renal artery** and the renal vein carries blood away. The liver has the **hepatic artery** to supply blood and the hepatic vein to return blood to the heart. Unusually the vein carrying blood away from the stomach and small intestine does not return it directly to the heart; it is carried first to the liver in the **hepatic portal vein**.

Hints and Tips

You have come across most of the blood vessel names and functions in Standard Grade. The three new names – the carotid artery, jugular vein and hepatic portal vein – feature frequently in the exam. It's worth the extra effort to learn their names and functions.

Cardiac cycle

Systole is the term that means the muscular wall of a heart chamber is contracting, whilst **diastole** is the term used to mean the heart chamber muscle is relaxing.

As can be seen from Figure 10.3 (1), when the atria fill with blood both the atria and ventricles are in diastole and the build-up of pressure causes the atrio-ventricular valves to open allowing blood to flow into the ventricles. In atrial systole (2) the walls of the atria contract, forcing the rest of the blood through the atrioventricular valves into the ventricles. Ventricular systole (3) is when the walls of a ventricle contract, closing the atrioventricular valves and forcing the blood up through the semilunar valves. At the start of ventricular diastole the pressure in the arteries becomes greater than in the ventricles and the semilunar valves close, thus preventing back flow. The 'lub-dub' sounds heard through a stethoscope are the atrioventricular and semilunar valves closing.

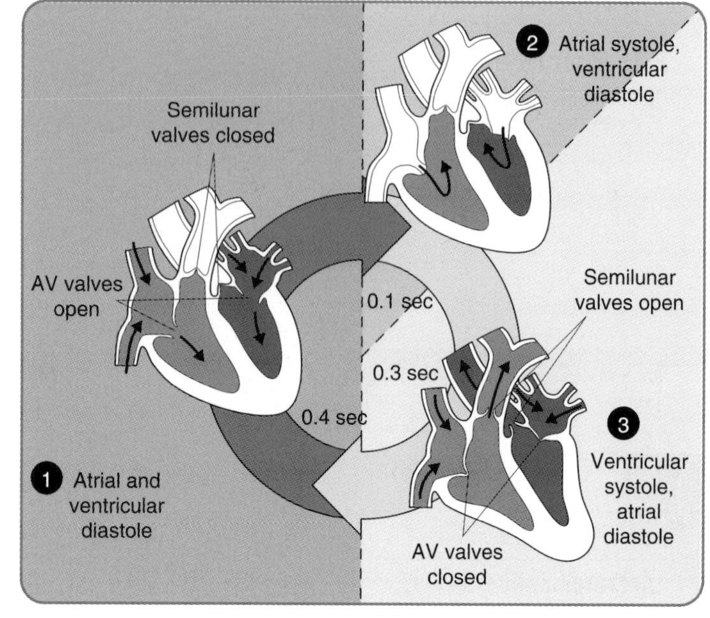

Figure 10.3 Cardiac cycle

For Practice

Complete a table like this to show if the two types of valve are opened or closed in each part of the cycle.

Stage in cardiac cycle	Atrioventricular valve	Semilunar valve
Diastole	Opens as blood fills atria	
Atrial systole		
Ventricular systole		

Control of the cardiac cycle

The **sinoatrial node** (**SAN**), situated in the wall of the right atrium (Figure 10.4), ensures that both atria contract simultaneously by sending out electrical impulses which are carried through the muscular walls. Impulses from the SAN reach the **atrioventricular node** (**AVN**) which in turn sends out impulses that are carried by conducting fibres to every part of the ventricular muscle. This causes the ventricles to contract simultaneously.

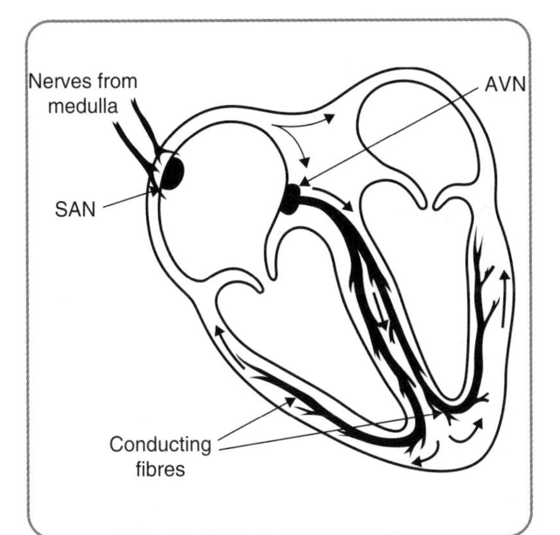

Figure 10.4 Control of cardiac cycle

Blood pressure

Blood pressure is highest in the aorta and lowest in the vena cava.

Pressure in the arteries reduces very little along their length as the elastic walls stretch during ventricular systole (the pulse) and spring back, maintaining nearly as high pressure, during diastole.

The greatest reduction in pressure is in the arterioles, as friction against the walls of the many narrow vessels resists the flow. The pressure gets less and less as the blood flows through the capillaries, then the venules and then the veins.

For Practice

What is the arterial blood pressure at systole and diastole?

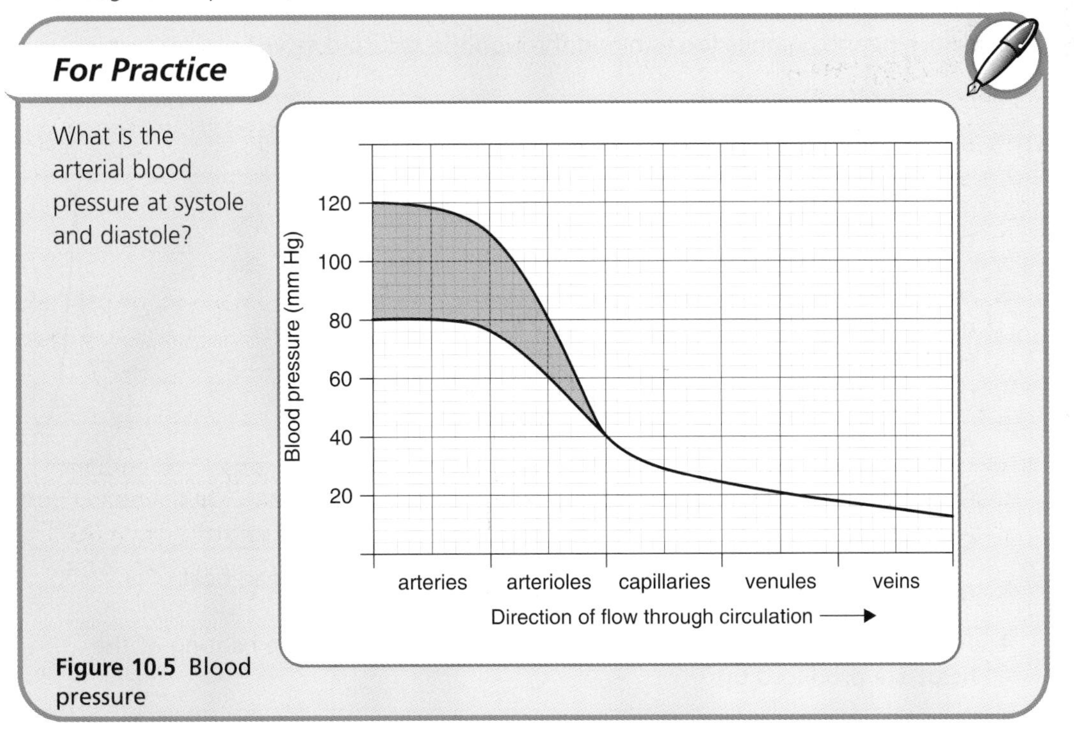

Figure 10.5 Blood pressure

Lymph circulation

Excess tissue fluid is returned to the heart in the lymphatic system which depends on valves and compression by skeletal muscles to force the lymph along. Lymphatic nodes contain a network of fibres that filter the lymph. The nodes contain **macrophages** which engulf and destroy micro-organisms by phagocytosis. The lymphatic system also transports the **products of fat digestion** absorbed by the lacteal in each villus.

For Practice

Make flashcards for all of the words that appear in **bold** type in the text of this chapter.

Exam Questions

1 Which row shows the changes in the blood as it moves from the arterioles through the capillaries to the venules?

	Oxygen	Carbon dioxide	Glucose	Pressure
A	increase	decrease	decrease	decrease
B	decrease	increase	decrease	decrease
C	decrease	decrease	increase	decrease
D	decrease	decrease	decrease	increase

2 Which heart chamber forces blood through the bicuspid valve?

A Left atrium

B Left ventricle

C Right atrium

D Right ventricle

3 Figure 10.6 shows the beat of a human heart recorded by an electrocardiogram. What is the heart rate in beats per minute (bpm)?

A 60 bpm

B 80 bpm

C 100 bpm

D 120 bpm

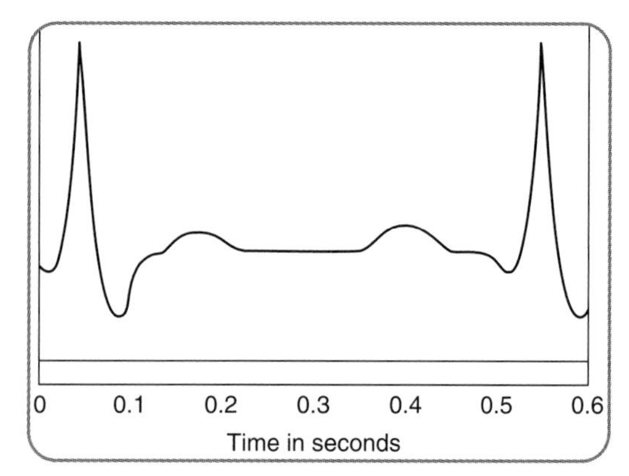

Figure 10.6 Human heart beat

4 The 'lub-dub' sounds heard through a stethoscope during the beating of the heart are produced by:

Exam Questions continued

A contraction of the ventricles

B contraction of the atria

C opening of the heart valves

D closing of the heart valves.

5 Which blood vessel carries blood to the head?

A Carotid artery

B Hepatic artery

C Jugular vein

D Pulmonary vein

6 Figure 10.7 shows pressure changes in the left atrium, left ventricle and aorta during a cardiac cycle.

(a) When did ventricular systole begin? *(1)*

(b) Name the valves that close at point E and at point I. *(2)*

(c) Between which two points would the pulse be felt? *(1)*

(d) Explain the rise in atrial pressure between points C and D. *(1)*

(e) Calculate the heart rate for this individual. *(1)*

7 Describe the roles of the SAN and AVN in the cardiac cycle. *(2)*

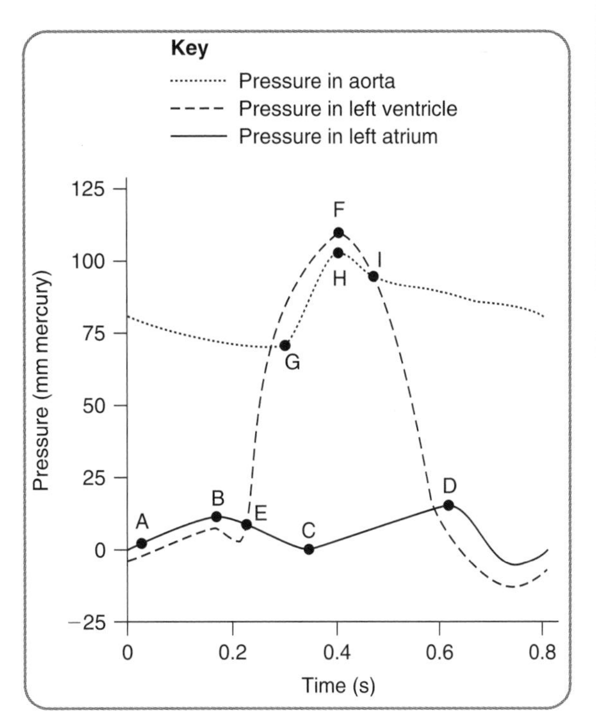

Figure 10.7 Pressure changes during cardiac cycle

Exam Questions continued

8 Figure 10.8 shows part of the lymphatic system.

(a) Describe how the flow of lymph fluid through the lymphatic vessels is maintained. *(2)*

(b) Lipids are absorbed into the lymphatic system in the small intestine. Describe how the structure of a villus is adapted for the efficient absorption of lipids. *(2)*

(c) State a function of the lymph nodes. *(1)*

9 Give an account of the pulmonary circulation (through heart and lungs) and the systemic circulation (through heart and the rest of the body). *(10)*

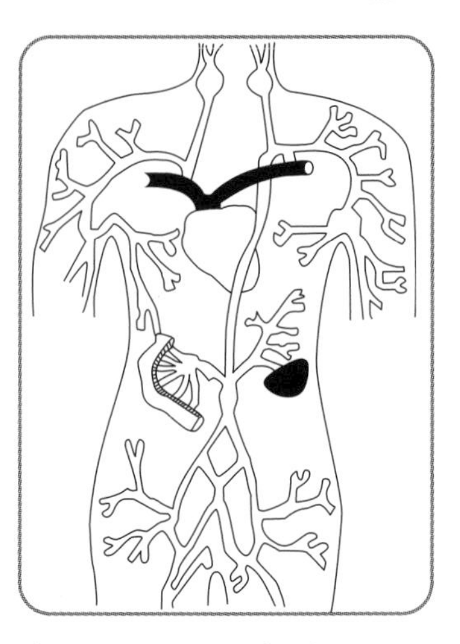

Figure 10.8 Lymphatic system

Answers

1 As blood travels through the tissues, oxygen and glucose are used by the cells and carbon dioxide is produced. In addition, pressure decreases from friction as blood is forced through the narrow capillaries. **Answer** B

2 **Answer** A

3 This frequently asked question puzzles many students. There is a trick here that is easily learned. First of all work out the time for one complete cycle by looking for two parts of the graph that look the same: maybe from 0 to 0.5 or 0.1 to 0.6 or even from the two peaks from 0.05 to 0.55. This gives a time of 0.5 s in this case. Then merely divide 60 s by the time for one beat. So $60 \div 0.5 = 120$. **Answer** D

4 The heart sounds heard through a stethoscope are caused by the atrioventricular (lub) and semilunar (dub) valves closing. **Answer** D

5 The carotid artery carries blood to the head. **Answer** A

6 (a) Ventricular systole is the contraction of the ventricle muscles which causes a sharp rise in pressure. This occurs at about 0.2 s.

(b) The rise in pressure in the ventricle at point E will cause the atrioventricular (bicuspid) valve to close to prevent back flow into the atrium. At point I, the pressure in the aorta becomes greater than in the ventricle and so the semilunar valve closes.

Answers continued

(c) The pulse will be felt between points G and H as there is a surge of pressure in the aorta. In truth, it will happen a fraction of a second later when the surge reaches the wrist or other pulse point.

(d) This is the part during atrial diastole when blood is forced in from the vena cava. As the atrium fills, the pressure increases.

(e) Seems to have gone through the whole cycle in 0.8 s. So the trick is to divide 60 s by 0.8 and that gives a heart beat of 75 bpm.

7 SAN sends out a wave of excitation that causes both the atria to contract simultaneously (1 mark).

AVN sends impulses down the conducting fibres to cause both ventricles to contract simultaneously (1 mark).

8 (a) Lymph forced along by surrounding skeletal muscle contractions (1 mark). Valves prevent back flow (1 mark).

(b) The villus has a lacteal to absorb and transport away lipids (1 mark). It also has a large surface area or a wall that is one cell thick (1 mark).

(c) The lymph nodes contain macrophages to engulf and destroy micro-organisms.

9

Nos	Mark scoring points		Comments
	Pulmonary circulation		*Write the sub-heading*
1	Blood returns from the body in the vena cava	1	
2	Collects in right atrium	1	
3	Right atrium pumps blood into the right ventricle	1	
4	Through the tricuspid valve	1	*Atrioventricular valve is a good answer here too. However, it counts only once later when you should have mentioned the bicuspid valve too.*
5	Right ventricle pumps blood to lungs in pulmonary artery	1	
			Maximum 5 marks
	Systemic circulation		*Write the sub-heading*
6	Pulmonary vein returns blood to heart	1	
7	To the left atrium	1	

Answers continued

Nos	Mark scoring points		Comments
	Pulmonary circulation		*Write the sub-heading*
8	Left atrium pumps blood to left ventricle	1	
9	Through bicuspid valve	1	*Atrioventricular valve is a good answer here too. However, it counts only once.*
10	Left ventricle pumps blood to body through the aorta	1	
			Maximum of 5 marks
	There are more marks available for either pulmonary circulation or systemic circulation if you mention the following.		
11	Valves prevent back flow	1	
12	Right ventricle produces less pressure	1	
13	Contraction of heart muscle is known as systole *or* relaxation is diastole	1	
14	Right side of heart has deoxygenated blood	1	*Or converse*
15	Arteries then arterioles then capillaries *or* capillaries then venules then veins	1	

Chapter 11

DELIVERY AND REMOVAL OF MATERIALS

Oxygen

When oxygen diffuses into a red blood cell it combines with **haemoglobin** to form oxyhaemoglobin. Figure 11.1 shows that haemoglobin becomes saturated with oxygen in the lungs where there is a high oxygen concentration. The lower oxygen concentration in the tissues of a resting person causes haemoglobin to unload some of its oxygen. The very low oxygen tension in exercising muscles causes haemoglobin to unload nearly all its oxygen. In addition, a raised temperature reduces haemoglobin's affinity further and so it unloads even more oxygen in muscles heated by strenuous activity.

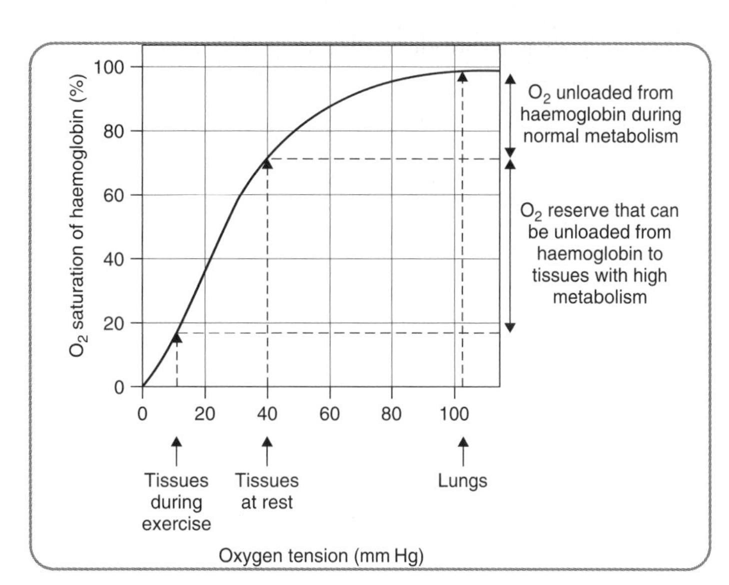

Figure 11.1 Haemoglobin/oxygen saturation

For Practice

1. From Figure 11.1, what percentage of the oxygen is unloaded by haemoglobin at the oxygen tension in muscles at rest and in exercising muscles?
2. How would the plot for haemoglobin affinity differ in muscles warmed by strenuous exercise?

Red blood cells are small and have a biconcave disc shape which gives them a large surface area to volume ratio for oxygen to diffuse in and out. Their shape ensures that no haemoglobin molecule within the cell is far from a surface. They do not have a nucleus or other cell organelles, being just a membrane container packed full of haemoglobin. They are flexible so they can pass through capillaries which have a diameter only slightly bigger than the cells.

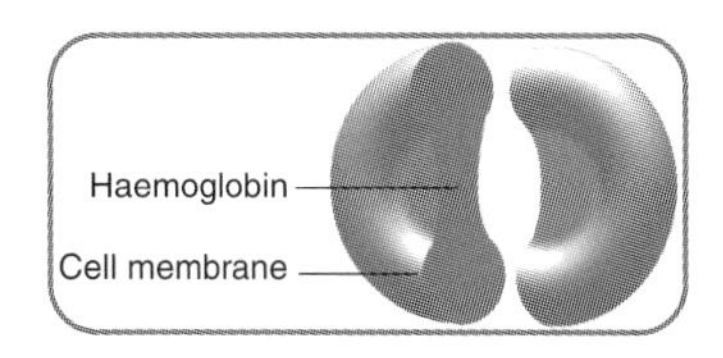

Figure 11.2 Red blood cells

Red blood cells are manufactured in the bone marrow using **iron** and **vitamin B_{12}** as essential raw materials. They survive for an average of 120 days and then are broken down in the spleen, liver and bone marrow. The haemoglobin is broken down to **bilirubin** and excreted in the bile and the iron is stored in the liver.

Absorption of nutrients

Bile is made in the liver and stored in the gall bladder and pours into the start of the small intestine. Bile emulsifies large fat globules into many tiny droplets, which gives lipase many more fat molecules on the surface to act on.

For Practice

1 Which would take longer to completely digest: the fat in large globules or in billions of tiny droplets?

2 Why might a gall bladder removal operation be an effective, if rather drastic, slimming aid?

The surface of the long small intestine is folded and covered in millions of villi – all combining to give a large surface area for molecules to be absorbed. Villi also have walls one cell thick to give a short diffusion pathway.

Glucose and amino acids diffuse into the blood capillaries. Fat digestion products, fatty acids and glycerol, are absorbed by the lacteal and transported by the lymphatic system. Vitamin B_{12}, needed for red blood cell manufacture, can only be absorbed by the villi if a special protein called an **intrinsic factor** is present.

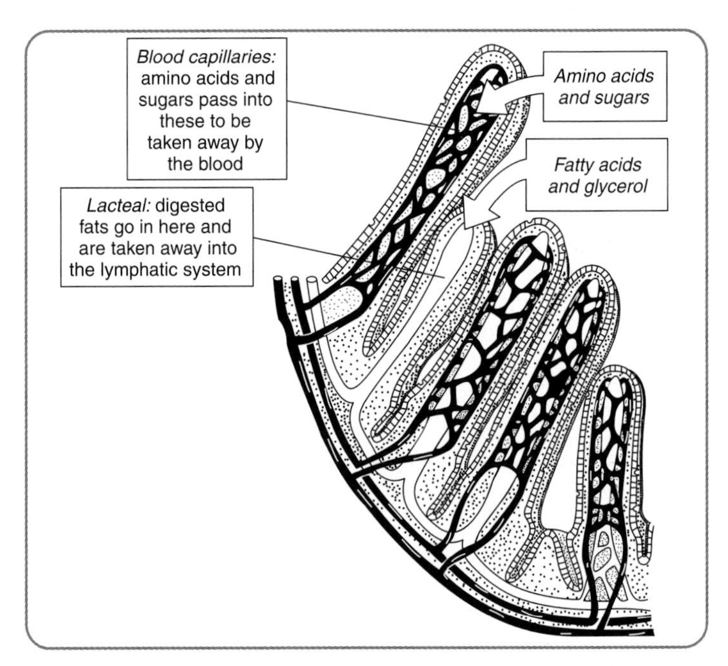

Figure 11.3 Villi in small intestine

Glucose is used by cells as a substrate for respiration, stored as liver and muscle glycogen, or it can be converted to fat for longer-term storage. Lipids may also be used as a substrate for respiration or stored around the body (also see Chapter 4). Amino acids diffuse into cells and are used in protein synthesis. Some vitamins and minerals act as enzyme activators.

Removal of materials from the blood

The role of the liver

The liver is supplied with oxygenated blood and the optimum concentration of numerous molecules directly from the heart by the hepatic artery. The liver is said to have a dual supply because it is also supplied with blood from the stomach and small intestine by the **hepatic portal vein**. As the blood in the hepatic portal vein has already been to an organ it is deoxygenated. After a meal the blood in the hepatic portal vein has a high concentration of absorbed substances such as glucose, amino acids, vitamins and minerals. The liver conserves useful substances by removing excess molecules from the blood and storing them until required. As a consequence, the deoxygenated blood in the hepatic vein has the optimum concentration of glucose and amino acids. Urea is made in the liver, so the hepatic vein often has a high urea concentration.

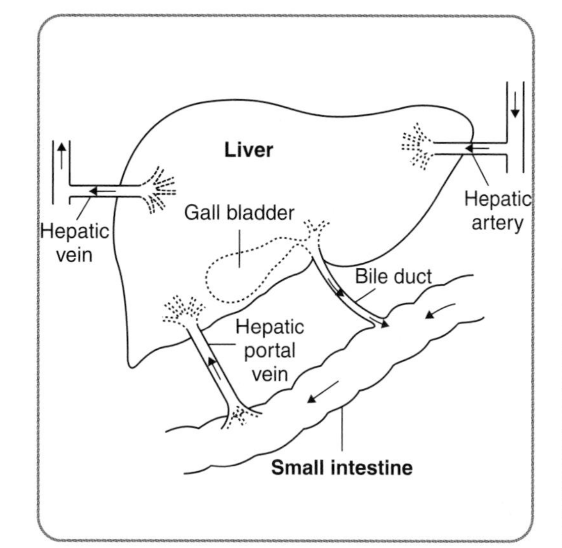

Figure 11.4 Structure of the liver

For Practice

1 Complete a table like the one below to help memorise the contents of the blood vessels going to and from the liver. Assume that a meal has recently been digested.

Molecules	Hepatic artery	Hepatic vein	Hepatic portal vein
Oxygen			
Glucose			
Amino acids			
Urea			

2 Which liver blood vessel has different molecule concentrations after meals and between meals?

The liver removes absorbed molecules brought in by the hepatic portal vein to maintain a constant blood concentration.

- Excess blood glucose is converted to **glycogen**. This will be changed back when blood glucose concentration goes below the norm.
- Excess amino acids are either changed into others or broken down by a process known as **deamination**. Deamination provides a substrate for respiration, and ammonia which is changed to the less toxic **urea**.

- Lipids are modified, some being converted to cholesterol which is required for membranes.
- Toxic materials, such as alcohol, are broken down by the liver.
- Bilirubin, the breakdown product of haemoglobin, is removed from the blood by the liver and excreted in the bile.
- Excess fat soluble vitamins A, D, E and K are also stored.

The role of the lungs

Carbon dioxide is produced by aerobic respiration in all living cells. It is carried to the lungs in the plasma. Carbon dioxide diffuses from the plasma into the alveoli down a concentration gradient. Regular breathing movements exhale carbon dioxide rich air and maintain the steep concentration gradient. At the same time oxygen diffuses from the high concentration in the alveoli into the red blood cell where it combines with haemoglobin.

> ***Hints** and **Tips***
>
> All body surfaces for absorption – alveoli, villi and capillaries – have a large surface area so that countless billions of molecules can pass through together every millisecond.

The role of the kidneys

The role of the kidneys is to remove urea and other toxic waste, and to keep the blood water concentration within fine limits. Blood rich in useful molecules and urea is brought to the kidney in the renal artery. Blood that is still rich in useful molecules but with urea removed and having an adjusted water concentration, leaves the kidney in the **renal vein**.

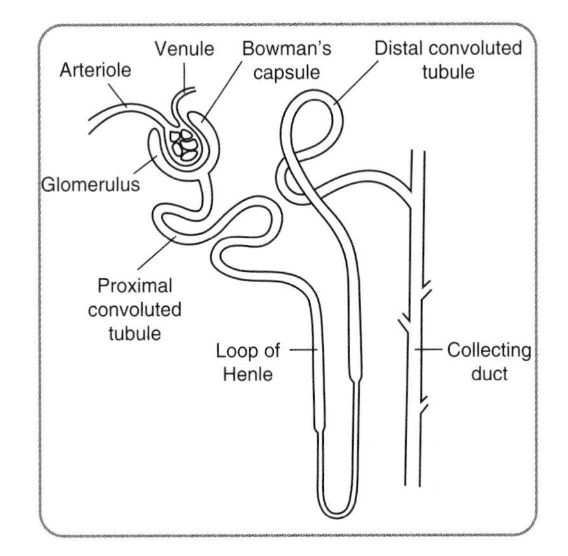

Figure 11.5 Kidney nephron

Blood from an arteriole enters a ball of capillaries, called the **glomerulus**, surrounded by the **Bowman's capsule** which is the start of a structure known as a **nephron**. Not all the blood entering the glomerulus can leave in the venule as it is narrower than the arteriole. Blood pressure and the 'traffic jam' forces plasma containing water, urea and useful molecules such as glucose out through pores in the glomerulus into the Bowman's capsule. Blood cells and large plasma protein molecules remain in the capillaries, so the blood is said to have been filtered – this is known as ultra-filtration. Useful molecules in the glomerular filtrate, such as glucose and amino acids, are reabsorbed by active transport in the next part of the nephron called the proximal convoluted tubule. Water diffuses out by osmosis. As the tubule fluid passes through the Loop of Henle, salts are pumped out by active transport to produce a low water concentration around the collecting duct. Some salt and water are reabsorbed from the distal convoluted tubule. The liquid flowing down the collecting duct – urine – contains mainly water, urea and excess salt.

Control of blood water concentration

The walls of the collecting duct are impermeable to water. Water cannot diffuse down the steep concentration gradient from the duct into the surrounding salty solution without **anti-diuretic hormone** (**ADH**). The permeability of the collecting duct is increased by ADH. The hormone ADH is always present and so reabsorption of water occurs constantly. Control of blood water concentration depends on increasing or decreasing the rate of absorption by changing the blood concentration of ADH.

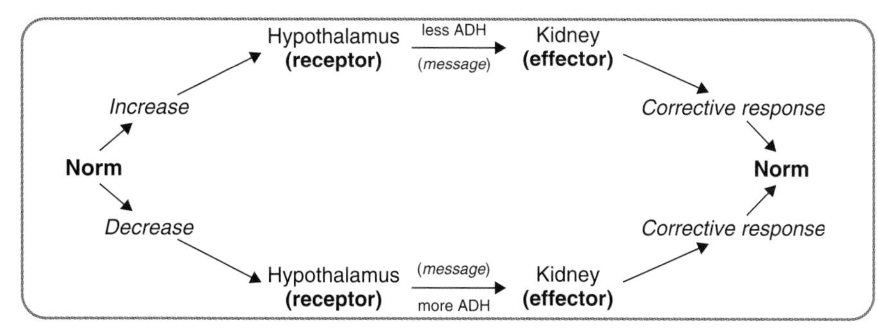

Figure 11.6 Control of blood water concentration

Hints and Tips

Often it helps to know what a biological word means. *Diuretic* in ADH means urine production. So *anti*-diuretic means less urine. More ADH, less urine. Less ADH, more urine.

As can be seen in Figure 11.6, changes in blood water concentration are detected by the **hypothalamus**. If the blood water concentration rises higher than the norm, the **pituitary gland** releases less ADH. The hormone is carried to the kidney in the blood. Less ADH causes the collecting duct wall to becomes less permeable to water and less water is reabsorbed. A large volume of dilute urine will be produced and the blood water concentration falls towards the norm. When the norm is detected by the hypothalamus, the ADH produced by the pituitary increases again. This is an example of negative feedback. As can be seen in Figure 11.6, the opposite occurs if the blood water concentration falls below the norm.

Hints and Tips

Have you noticed that every time a hormone with a long name is mentioned, often shortened to letters, it is produced by the pituitary gland? Remember that fact and you will not have to learn the separate facts that ADH and FSH and LH (sometimes called ICSH in males) are all made by the pituitary gland. All have the word *hormone* at the end, too.

For Practice

Make flashcards for all of the words that appear in **bold** type in the text of this chapter.

Exam Questions

1 Which of the following are needed for haemoglobin synthesis?

A Calcium and intrinsic factor

B Iron and intrinsic factor

C Calcium and vitamin B_{12}

D Iron and vitamin B_{12}

2 Figure 11.7 shows blood circulation through the liver.

After a meal high in protein and starch, which of the following have a lower concentration in blood vessel K than in blood vessel L?

A Amino acids and glucose

B Urea and glucose

C Amino acids and glycogen

D Urea and fatty acids

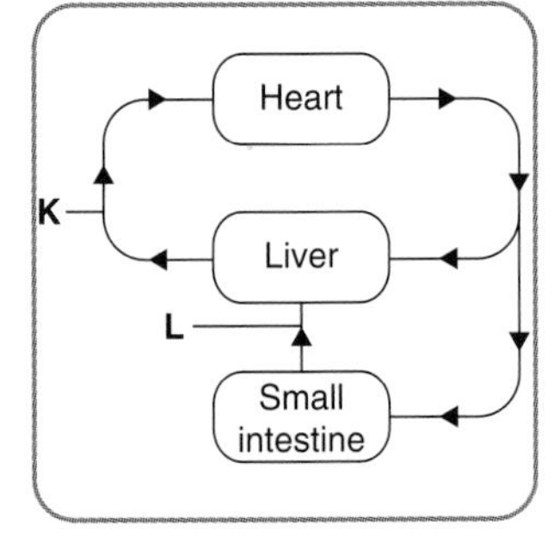

Figure 11.7 Blood circulation in liver

3 The table below shows the changes in concentration of sodium and urea as they pass through the kidneys of a human.

Substance	Mass filtered (g/24 hours)	% of each substance reabsorbed
Sodium	600	99.5
Urea	50	40.0

The ratio of sodium to urea in the urine, expressed as the simplest whole number ratio, is:

A 1 : 10

B 1 : 12

C 1 : 100

D 1 : 120.

Exam Questions continued

4 (a) Figure 11.8 shows the effect of temperature on oxygen affinity of haemoglobin at different oxygen tensions.

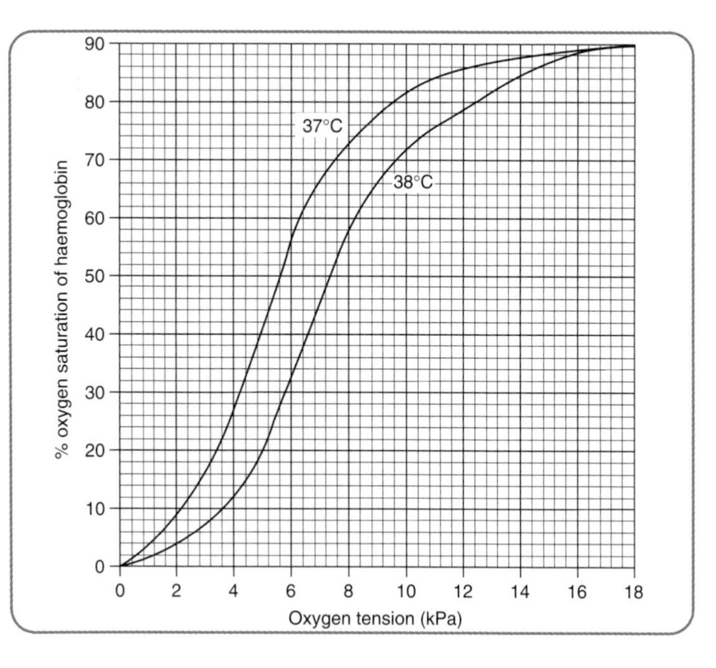

Figure 11.8 Oxygen tension

(i) For these two temperatures, at which oxygen tension is the difference in the percentage saturation of haemoglobin with oxygen greatest? *(1)*

(ii) Explain the effect of vigorous exercise on the oxygen supply to muscle tissue. *(2)*

(b) The information refers to stages in the life cycle of a red blood cell.

Stage 1 Red blood cell produced

Stage 2 Red blood cell functions for 120 days

Stage 3 Red blood cell broken down

Stage 4 Fate of breakdown products

(i) Name the site of red blood cell production referred to in Stage 1. *(1)*

(ii) Name two organs in which red blood cells are broken down. *(1)*

(iii) Iron released from the breakdown of red blood cells can be stored and used later in the production of new red blood cells. Describe the fate of another product of the breakdown of red blood cells. *(2)*

Exam Questions continued

5 Figure 11.9 represents the control system for water concentration of the blood.

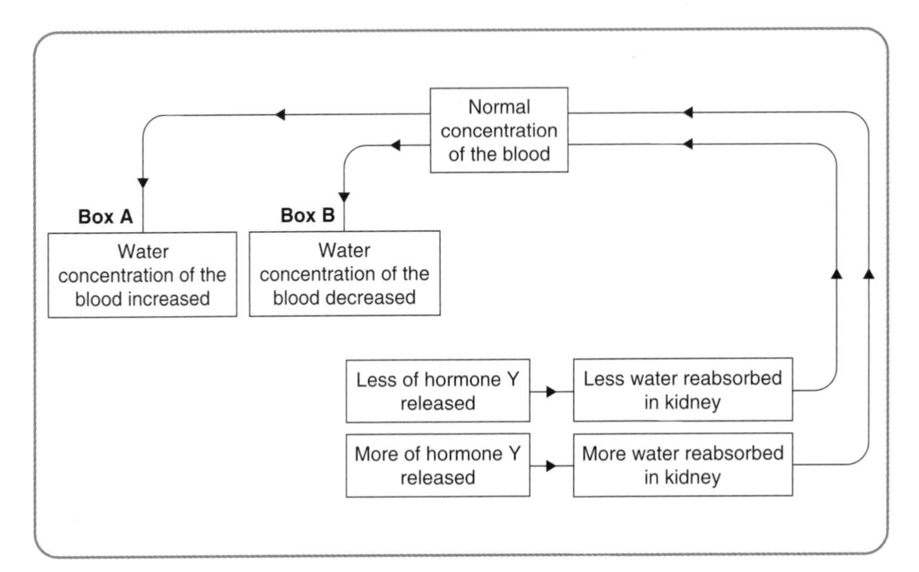

Figure 11.9 Water concentration of blood

(a) Draw and complete a similar diagram by connecting boxes A and B to the appropriate hormone response. *(1)*

(b) Name hormone Y. *(1)*

(c) Explain how control of water concentration of the blood illustrates the principle of negative feedback. *(1)*

(d) If the water concentration of the blood remained lower than normal due to failure of the control system, what effect would this have on blood cells? *(1)*

6 Figure 11.10 shows a kidney nephron.

(a) Name parts A, B and C. *(2)*

(b) Describe the process of ultrafiltration. *(2)*

(c) Give an example of a molecule reabsorbed at A. *(1)*

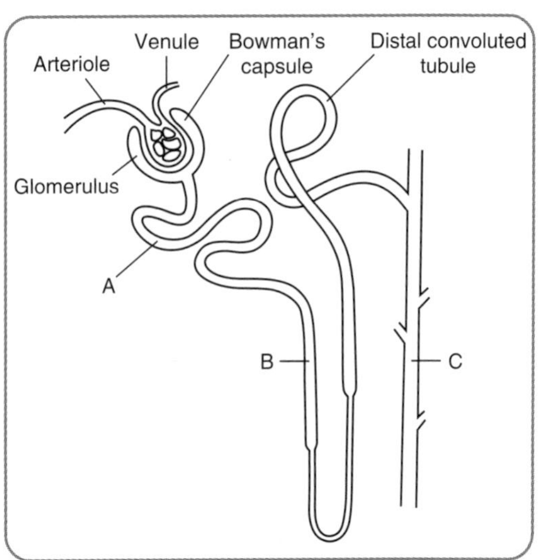

Figure 11.10 Kidney nephron

Answers

1 Iron and vitamin B_{12} are needed for haemoglobin synthesis. **Answer** D.

2 After a meal of protein and starch the digestion products of amino acids and glucose are absorbed by the villi in the small intestine and carried to the liver by the hepatic portal vein. Thus vessel L has a high concentration of amino acids and glucose. In the liver excess glucose is converted to glycogen and excess amino acids are deaminated, so vessel K will have lower concentrations of glucose and amino acids. You should have been able to discount the other options because: urea is made in the liver so will be higher in K; glycogen is too large to leave liver cells so will not be found in either vessel; fatty acids are a product of fat digestion and they are carried by the lymphatic system. **Answer** A

3 This is a tough one because it is a percentage and a ratio question. First, work out the mass of each in the urine. 99.5% of the sodium is reabsorbed so 0.5% will be in the urine. Easiest way is to first divide 600 by 100 to find 1% and then you easily calculate any percentage. 600 ÷ 100 = 6. So 0.5% is 3. Then, 40% of the urea is reabsorbed so 60% is in the urine. Again, find 1% by 50 ÷ 100 = 0.5. So 60 x 0.5 = 30.

Now you have the values of 3 for sodium and 30 for urea – a whole number ratio of 3 : 30, but not at its simplest. Divide both sides by 3 to get a ratio of 1 : 10. **Answer** A

4 (a) (i) You need to count the number of little boxes between each plot for each oxygen tension. At 6 kPa the number of little boxes is 12, or 24%. This seems more than at any other oxygen tension.

(ii) Vigorous exercise will increase the temperature of the muscles (1 mark). As the graph shows, at a higher temperature heamoglobin unloads more oxygen (1 mark).

(b) (i) Red blood cells are manufactured in the red bone marrow.

(ii) Red blood cells are broken down in the liver, spleen and bone marrow.

(iii) Once the iron is removed from the haemoglobin, the remainder is converted to bilirubin (1 mark). Bilirubin is excreted in the bile (1 mark).

5 (a) In the completed diagram, box A (blood water concentration increased) should be connected to less hormone produced *and* box B to more hormone produced.

(b) The hormone involved in blood water concentration control is ADH.

(c) When the blood water concentration returns to normal ADH is returned to the normal level or the corrective mechanism is switched off.

(d) The blood cells would shrink as they would lose water by osmosis.

6 (a) A = proximal convoluted tubule, B = Loop of Henle, C = collecting duct. Usually you get 2 marks for getting all 3 correct, and 1 mark for getting 1 or 2 correct.

(b) High pressure in glomerulus as the blood vessel entering is wider than the one leaving it, (1 mark) forcing filtrate into the Bowman's capsule (1 mark).

(c) Glucose and amino acids are reabsorbed at A by active transport. Water leaves too by osmosis.

Chapter 12

REGULATING MECHANISMS

Key Ideas

Negative feedback keeps factors like body temperature or the concentration of many different molecules close to their optimum. When a factor increases or decreases from the norm, the change is sensed by a receptor. The receptor then sends a message to an effector in another part of the body. The message may be a hormone or a nerve impulse. The effector responds to the message by switching on a corrective response that returns the factor to the norm again. When the receptor detects that the factor has returned to the norm it stops sending messages to the effector and the corrective response is switched off.

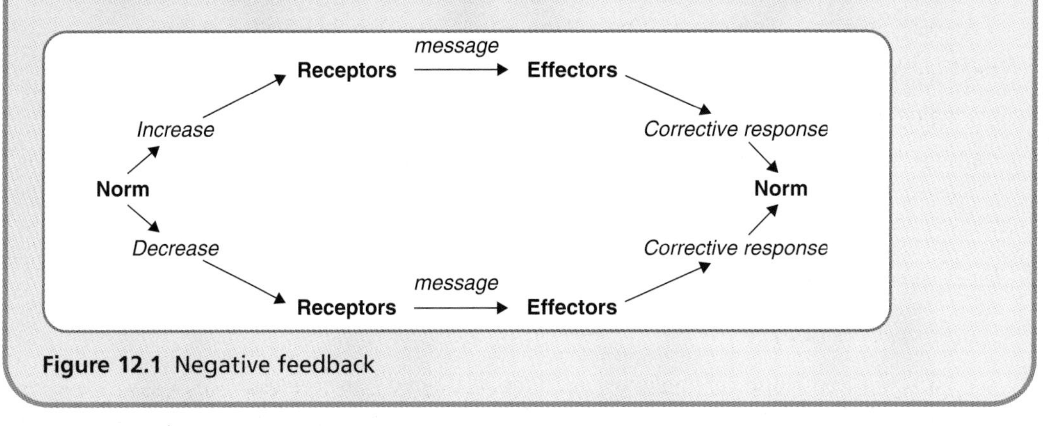

Figure 12.1 Negative feedback

The control of heart rate

The **sino-atrial node** (SAN) maintains a regular, synchronised heart beat all by itself. However, this heart rate is not fast enough for strenuous activity and too fast for calm rest. A part of the brain called the **medulla** adjusts the heart rate to match the activity level of the body. The medulla is connected to the SAN by two nerves that are branches of the autonomic nervous system (see Chapter 13). Impulses in the **sympathetic nerve** increase the heart rate in response to higher blood carbon dioxide when the body is active. **Parasympathetic nerve** impulses slow the heart rate in response to lower

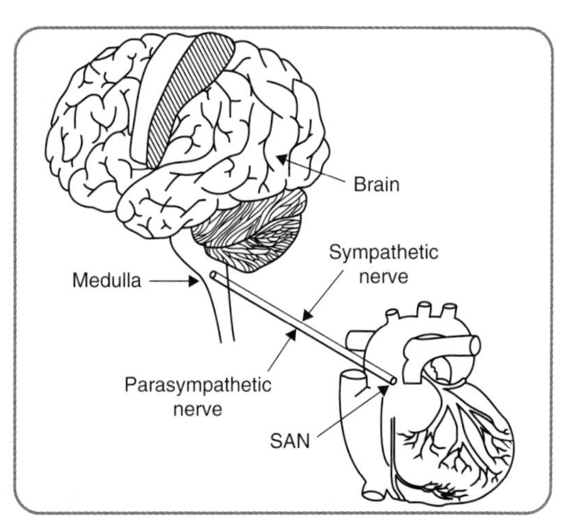

Figure 12.2 Control of heart rate

blood carbon dioxide during quiet periods. When the medulla senses that the blood carbon dioxide concentration has returned to the norm it stops sending impulses to the SAN and the corrective response is switched off.

For Practice

Complete a table like the one below to show that the control of heart rate is an example of negative feedback.

Activity level (blood CO_2 concentration)	Receptor	Message	Effector	Corrective response
Increase				
Decrease				

The effect of exercise on the cardiovascular and respiratory systems

Exercising muscles require more ATP which they get from an increased rate of respiration. Muscle cells with a higher respiration need more glucose and oxygen and also produce more carbon dioxide. The rise in carbon dioxide concentration is detected in receptors in the medulla which responds by sending impulses along the sympathetic nerve system.

Sympathetic nerve stimulation increases the heart and breathing rate. As can be seen in the table below, the volume of blood pumped per beat and the volume of air inhaled in each breath is also increased.

Activity level	Heart rate (beats/min)	Stroke volume (cm^3)	Breathing rate (breaths/min)	Volume per breath (l)
Rest	72	70	12	0.5
Exercising	120	112	24	0.9

A greater cardiac output from a faster heart rate with a greater blood volume per stroke, transports more glucose, oxygen and carbon dioxide. Faster, deeper breathing supplies more oxygen and removes more carbon dioxide.

For Practice

Calculate the cardiac output (heart rate × stroke volume) and volume of air inhaled per minute (breathing rate × volume per breath) before and during exercise to appreciate the dramatic effect.

Part of body	Blood volume At rest (cm^3/min)	During exercise (cm^3/min)
Brain	850	850
Muscles	1,200	12,500
Skin	500	2,000
Intestines	1,500	500
Kidney	1,100	600
Heart	250	750
Other	600	400

During exercise, sympathetic nerve impulses cause a redistribution of blood to the tissues. As can be seen in the table, much more blood goes to the muscles to provide extra food and oxygen and to remove more carbon dioxide. This includes the heart muscles. The blood flow to the skin is also increased so that the heat from rapidly contracting muscles can be lost.

The muscular walls of the arterioles respond to sympathetic nerve stimulation by relaxing, which makes them wider. This is known as **vasodilation**. The opposite occurs in parts of the body that have nothing to contribute to strenuous activity. **Vasoconstriction** narrows the arterioles and less blood goes to organs, such as the intestines and the kidneys, during exercise.

The heart rate and breathing rate are also under the control of the hormone adrenaline. At times of strenuous activity, a sympathetic nerve stimulates the adrenal glands to release adrenaline which causes greater activity in the intercostal and diaphragm muscles and in the heart SAN.

Hints and Tips

The actions of the sympathetic and parasympathetic nervous systems are frequent exam questions. Remember:

- **Sympathetic** = action – more blood to muscles and skin and speeds up activity of heart and lungs; slows digestion and excretion.
- **Parasympathetic** = calm – more blood to digestive system and kidneys; slows the heart and breathing rate.

Blood sugar

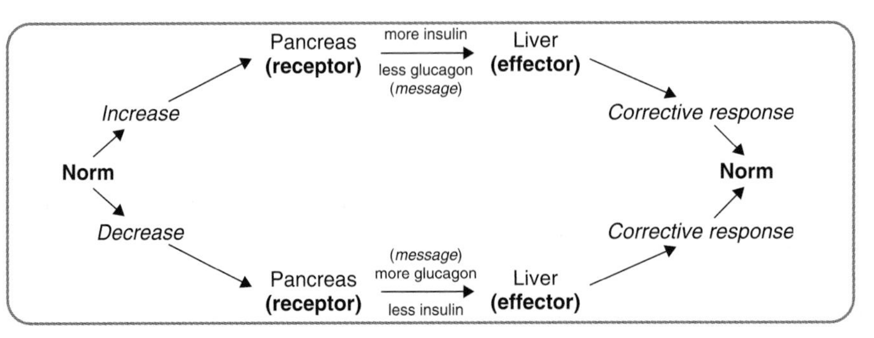

Figure 12.3 Control of blood sugar

When the blood glucose concentration *increases*:

- the increase is detected by the pancreas
- the pancreas releases more insulin and less glucagon
- the two hormones travel to the liver in the blood
- a high concentration of insulin makes the liver cells absorb more glucose
- in the liver, glucose molecules are joined together to form **glycogen**.

With glucose removed, the blood glucose concentration decreases to the norm. This is sensed by the pancreas which reduces the insulin output.

When the blood glucose concentration *decreases*:

- the decrease is detected by the pancreas
- the pancreas releases more glucagon and less insulin
- the two hormones travel to the liver in the blood
- the high concentration of glucagon causes glycogen to be broken down to glucose in liver cells.

Fresh supplies of glucose bring the blood glucose concentration back up to normal. This is sensed by the pancreas which reduces the glucagon output.

Hints and Tips

For some reason, hardly anyone remembers glucagon and when it is produced. But not you. Just remember: 'glucose has gone = glucagon'.

Adrenaline

At times of stress or strenuous activity, adrenaline is released. Adrenaline causes glycogen to be converted to glucose and inhibits the effect of insulin. Adrenaline pushes the blood glucose concentration to a higher than normal level, preparing for 'fight or flight'. Once the emergency is over, adrenaline levels fall and the effect of insulin brings the blood glucose concentration back down to normal.

For Practice

Complete a table like the one below to show that the control of blood glucose concentration is an example of negative feedback.

Blood glucose concentration	Receptor	Message	Effector	Corrective response
Increase				
Decrease				

Temperature

Body temperature is maintained close to 37°C. Enzymes work most efficiently at this optimum temperature. If the body temperature was lower than the optimum, insufficient energy would be released for survival. Higher temperatures may denature enzymes.

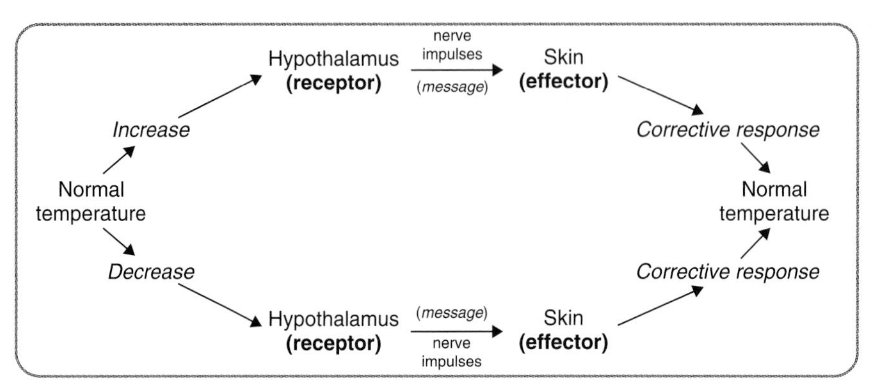

Figure 12.4 Control of body temperature

Body temperature is monitored by the **hypothalamus**. If the hypothalamus senses a change in blood temperature it sends nerve impulses to effectors in the skin that switch on corrective responses. When the hypothalamus senses the blood temperature has returned to normal it stops sending nerve impulses to the skin effectors and the corrective responses are switched off.

When the blood temperature *decreases* the hypothalamus switches on the following corrective responses:

- repeated contraction of skeletal muscles called **shivering** produces heat
- skin arterioles constrict (narrow) so less warm blood flows in the skin capillaries – this is called vasoconstriction, which reduces the heat lost by radiation
- **hair erector muscles** contract, raising the hair which traps air providing insulation
- release of the hormone **adrenaline** increases the metabolic rate, generating heat
- less sweat is produced.

Once the blood temperature is raised to the optimum this is detected by the hypothalamus, the stimulation of the effectors stops and the corrective responses are switched off.

When the blood temperature *increases* the hypothalamus switches on the following corrective responses:

- more **sweat** is produced and heat energy needed to evaporate sweat leaves the skin with the water molecules
- skin arterioles widen bringing more blood close to the surface (vasodilation) which allows more heat to radiate away
- hair remains flat, there is no shivering and adrenaline is not released.

Once the blood temperature is lowered to the optimum this is detected by the hypothalamus, the stimulation of the effectors stops and the corrective responses are switched off.

People also have voluntary responses to blood temperature changes. They are able to sit closer to a fire or put on another jumper when they feel cold. Alternatively, a swim or an ice cold drink are sensible voluntary responses to a raised blood temperature.

For Practice

Complete a table like this to help memorise the corrective responses involved in temperature regulation. Tick the correct response.

Corrective response	Response to a blood temperature increase	Response to a blood temperature decrease
Shivering		
Sweating		
Vasodilation		
Vasoconstriction		
Hair muscle contraction		
Adrenaline production		

Hypothermia

Hypothermia occurs when the control by the hypothalamus is not able to replace heat as fast as it is lost. Infants may lose heat faster than they generate it because their surface area to volume ratio is so large. Each cell in an infant has more skin to lose heat through than a cell in an adult. Also their involuntary temperature control system is not fully developed. Parents have to take care that their infants are clothed sufficiently.

In extreme weather conditions adults may also experience **hypothermia**, as a result, for example, of wearing wet clothes in a blizzard. This results in a heat loss which is so fast that it cannot be countered by the hypothalamus and the effectors.

The **metabolic rate** of older people is lower, and their temperature control system is not as efficient. Unfortunately, whilst their voluntary control might suggest the need for a warmer home, the cost may not be affordable. Many regard it a disgrace that in a rich, 'caring' UK society, thousands of older people die as a result of hypothermia every winter.

Hints and Tips

Sometimes an exam question asks you to work out the surface area to volume ratio from a diagram like Figure 12.5.

Every cube has a volume of $1 \times 1 \times 1 = 1^3$. In this case 16×1^3. Every cube has 6 sides, but all of them have at least one covered by another. Count and record the number of sides available to lose heat for each cube. Any side covered by another cube does not count; in this case, 60×1^2. So the SA to V ratio is 60 : 16. Trying for the simplest whole number ratio, this divides down to 30 : 8, then 15 : 4, but can go no further. So 15 : 4 is the best answer.

Figure 12.5 Surface area to volume

For Practice

Make flashcards for all of the words that appear in **bold** type in the text of this chapter.

Exam Questions

1 The table below shows changes in breathing before and after exercise.

Activity	Average volume of each breath (cm^3)	Breathing rate (breaths per minute)
Resting	500	10
Exercising	1500	20

The percentage increase in volume of air breathed per minute when the subject changed from resting to exercising was:

A 50% C 500%

B 100% D 5000%.

Exam Questions continued

2 Which of the following graphs shows the effect of exercise on the rate of blood flow to the brain?

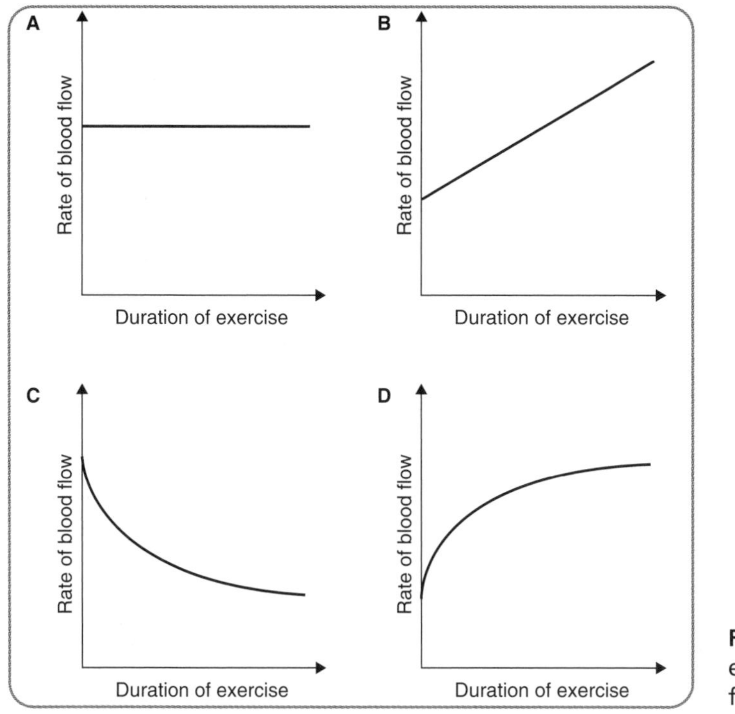

Figure 12.6 Effect of exercise on blood flow to the brain

3 Which line in the table describes control of human body temperature?

	Monitoring centre	Method of communication to effectors in skin
A	hypothalamus	hormonal
B	pituitary gland	nervous
C	hypothalamus	nervous
D	pituitary gland	hormonal

4 The following list refers to temperature regulation.

1 Vasoconstriction

2 Vasodilation

3 Increased sweating

4 Decreased sweating

5 Hairs raised

6 Hairs remain flattened

Exam Questions continued

Which of the responses counteract a decrease in body temperature?

A 2, 3 and 6

B 2, 4 and 5

C 1, 4 and 5

D 1, 3 and 6

5 Figure 12.7 shows the results of an investigation into the effect of exercise on heart rate and stroke volume of an untrained individual. Stroke volume is the volume of blood pumped from the heart in each beat.

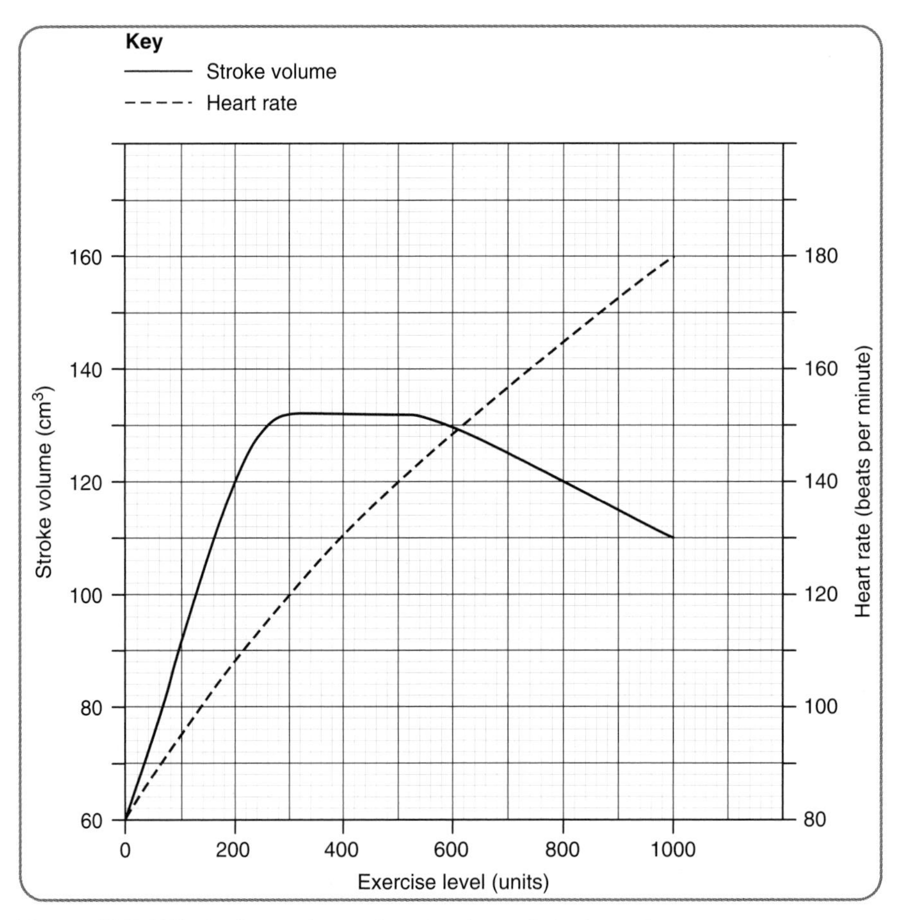

Figure 12.7 Effect of exercise on heart rate and stroke volume

(a) From the graph, what is the stroke volume when the heart rate is 120 beats per minute? *(1)*

(b) Use the values from the graph to describe changes in the stroke volume with increase in exercise level. *(2)*

Exam Questions continued

(c) Cardiac output is calculated from the following equation:

Cardiac output = heart rate × stroke volume

How do the results from exercise levels 800 and 1000 units support the statement that cardiac output remained relatively unchanged during this increase in exercise level for the untrained individual? *(1)*

(d) State the role of the sino-atrial node in the cardiac cycle. *(1)*

6 One form of diabetes is caused by failure to produce sufficient insulin. Figure 12.8 shows the results of tests on patients using insulin products from two manufacturers.

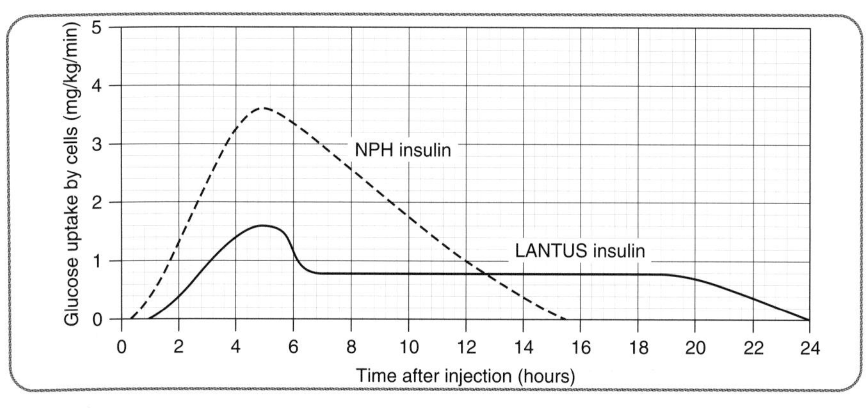

Figure 12.8 Insulin test results

(a) Insulin is measured as being effective when the glucose uptake by cells is above 0.4 mg/kg/minute. From the graph, calculate the difference in the time of effectiveness between the two insulin products. *(1)*

(b) Name the site of insulin production in the human body. *(1)*

(c) What happens to the glucose that is taken up by liver cells? *(1)*

(d) Name two other hormones that affect blood glucose concentration. *(2)*

7 Give an account of the mechanisms and importance of temperature regulation. *(10)*

One mark is available for coherence and one mark for relevance.

Answers

1 This is quite a difficult question as students are often not keen on percentage increase questions – but they are answerable, if you learn the simple trick.

First you need to work out the volume breathed per minute for each activity level by multiplying the average volume of each breath by the breathing rate. For resting it is $500 \times 10 = 5000$. For exercising it is $1500 \times 20 = 30\,000$.

Then the percentage increase. Work out the increase first: $30\,000 - 5000 = 25\,000$.

Then calculate the percentage increase by **dividing the increase by the starting number**. That is what you have to do every time. In this case, 25 000 divided by a resting volume of 5000 = 5. Then, because it is a percentage sum, multiply by 100 = 500%. **Answer** C

2 The brain requires a constant blood volume when the body is active or resting. The volume never changes. **Answer** A

3 The receptor for temperature control is the hypothalamus. The hypothalamus communicates with the skin by nerves. **Answer** C

4 A decrease in body temperature needs to be countered by corrective responses that conserve or produce heat. Vasoconstriction is the narrowing of skin surface arterioles and so less blood is brought close to the skin surface. Less heat is radiated away (1). Sweating loses heat by evaporating water and so heat is conserved by decreased sweating (4). Heat can be conserved by a layer of air trapped within raised hairs for insulation (5). **Answer** C

5 (a) With two vertical axes and two plots there are lots of ways of making an error with this sort of question. Heart rate is the axis on the right-hand side – it is not too difficult to find 120 beats. Follow the 120 beats line across to the heart rate plot which is the dotted line. They cut at 300 exercise units. Follow the 300 exercise units line up until it meets the stroke volume solid line plot. Then move across to the left-hand stroke volume axis to read off the value. There is another possible mistake opportunity here as many candidates think that every little box in a graph is worth 1 – and it hardly ever is. As there are 10 little boxes from 120 to 140, each little box is worth 2. So the answer is 132 cm^3.

(b) You should give the value every time the plot changes direction. It increases from 60 cm^3 to 132 cm^3, levels off between 300 and 500 exercise units then decreases to 110 cm^3. Make sure you give the unit or you will throw away a hard-earned mark.

(c) If you do the sums you will find that the cardiac outputs are the same. At 800, $165 \times 120 = 19\,800$ cm^3 per minute and for 1000, $180 \times 110 = 19\,800$ cm^3 per minute. As the question says 'how do the results … support the statement', you have to show your working for the mark.

(d) The SAN controls the heart rate *or* The SAN synchronises the contraction of the atria.

Answers continued

6 (a) First of all, decide what each little box represents. Five little boxes = 1, so each little box is worth 0.2. You need to find out where the two plots cut the line two little boxes up. The dotted NPH plot cuts it at 1 and at 14 = 13 hours. The solid Lantus plot cuts it at 2 and at 22 = 20. So the answer is 7 hours.

(b) Insulin is produced in the pancreas.

(c) Glucose taken up by liver cells is converted to glycogen.

(d) The hormones glucagon and adrenaline both cause liver cells to convert glycogen to glucose.

7

Nos	Mark scoring points		Comments
	Mechanisms		*Write the sub-heading. It helps you organise your answer, and is needed for the Cohesion mark*
1	Temperature regulation controlled by negative feedback	1	
2	Hypothalamus monitors the blood temperature	1	
3	Messages sent to skin via nerves	1	
4	Vasodilation in response to rise in temperature	1	*Could have got the same mark for the converse – vasoconstriction in response to a drop in temperature*
5	More heat lost by radiation	1	*Less heat lost by radiation*
6	Sweating in response to rise in temperature	1	*Or converse*
7	Heat lost by evaporation of water	1	*Or converse*
8	In cold conditions, hair erector muscles contract	1	*Or converse*
9	Trapped air gives better insulation	1	*Or converse*
10	Shivering in response to temperature drop	1	*Or converse*
11	Increased adrenaline in response to temperature drop	1	*Decreased adrenaline in response to temperature rise*
12	Increase in metabolic rate in response to temperature drop	1	*Decrease in metabolic rate in response to temperature rise*
			Any of the above for a ***maximum of 7 marks***

Answers continued

Nos	Mark scoring points		Comments
	Importance		*Write the sub-heading. It helps you to organise your answer, and is needed for the Cohesion mark*
13	Metabolism is controlled by enzymes	1	
14	Normal body temperature provides optimum temperature for enzymes	1	
			*Either of the above for a **maximum of 1 mark***

Coherence 1 mark if the information is grouped together under headings and at least 5 marks are gained.

Relevance 1 mark gained if at least 5 marks are gained and if you do not stray into other topics such as regulation of sugar or water concentration.

Chapter 13

THE NERVOUS SYSTEM

The brain

Evolution has greatly increased the size and capacity of the human brain. Its large size is the main difference between the human race and all the other animals. The colossal impact that people have had on the world is largely due to the thinking and problem solving skills which this allows.

The **cerebrum** is the part of the brain responsible for conscious thought. It is highly convoluted, which makes room for far more cell bodies to be located on the surface. This in turn allows many more possible interconnections to be made between neurones.

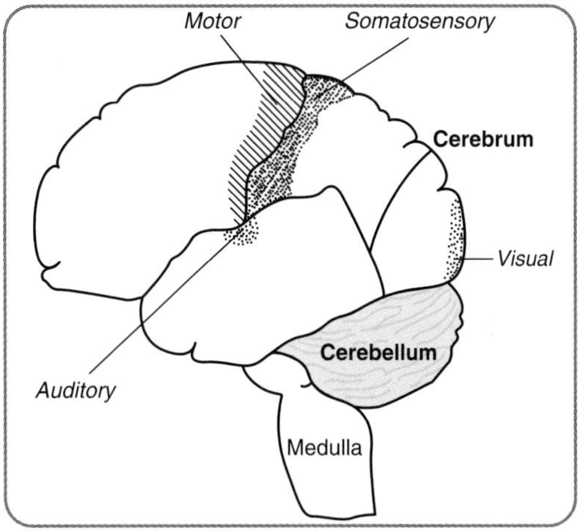

Figure 13.1 Structure of the brain

Different parts of the cerebrum are devoted to different functions – Figure 13.1 shows some of these. The **somatosensory area** receives impulses from the sensory receptors of the body. The **visual area** receives impulses from the eyes and the **auditory area** receives impulses from the ears. The **motor area** sends impulses to skeletal muscles to cause the appropriate movement. The **language areas** include various parts of the cerebrum responsible for communicating by speech, such as vocabulary memory, lip, tongue and vocal cord control and so on. **Association areas** close to each sensory area analyse incoming sensory impulses in the light of learning and experience.

Each area is linked to other areas to integrate all the incoming information and also the response.

The cerebrum is divided into two separate cerebral hemispheres, and information can be exchanged between them only through the area of tissue linking them – the **corpus callosum**.

A very large proportion of the motor area is concerned with controlling the thumb and fingers, giving fine motor control. The unequal distribution of motor areas in the brain to favour the delicate movements required for speech and for our incredible dexterity is illustrated by the model (Figure 13.2) of what a person would look like if the parts of the body were sized in proportion to the volume of brain controlling them. Note which parts of the body have the greatest control.

Figure 13.2 Model of relative parts of motor areas in the brain

Hints and Tips

You need to know the parts of the brain and the areas of the cerebrum, but you don't need to learn the exact positions of the areas of the cerebrum.

For Practice

1. What is the most significant difference between humans and all other animals?
2. Which part of the brain is responsible for conscious thought?
3. What is the advantage of the brain's convoluted surface?
4. Name six discrete areas of the cerebrum (note: there is no need to learn their positions).
5. Name three parts of the body which have an extra large part of the motor area to control them.

Organisation of the nervous system

As shown in Figure 13.3, the nervous system is divided into the **central nervous system (CNS)** and the **peripheral nervous system**. The CNS is composed of the brain and the spinal cord; the peripheral nervous system is made up from **sensory neurones** carrying impulses from the receptors to the CNS and **motor neurones** which carry impulses from the CNS to the muscles and glands.

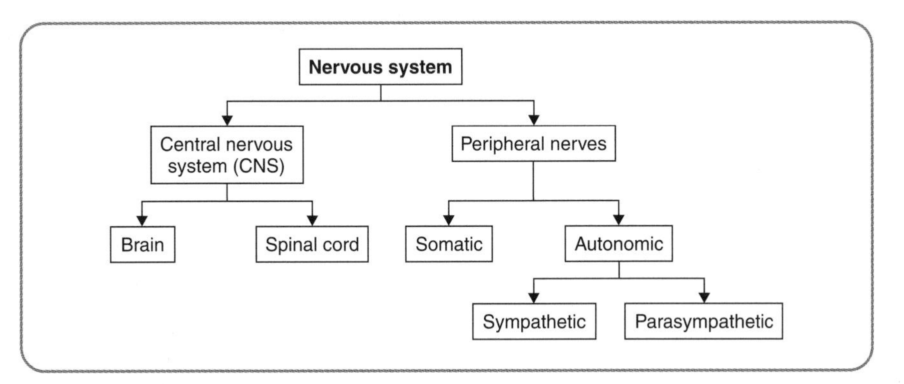

Figure 13.3 The nervous system

The peripheral nervous system consists of the **somatic** and the **autonomic** nervous systems. The somatic deals with functions under conscious control, such as walking and speaking, whilst the autonomic system controls functions that happen automatically, such as heart rate.

The autonomic system is divided into two parts, one for slowing things down and conserving resources and the other for raising activity levels. The **parasympathetic** part of the autonomic system is in control at times of rest and relaxation when the heart rate is low and blood flow is reduced in the muscles and increased to the digestive system. The **sympathetic** part of the autonomic system takes over when the body is active, or excited, and causes increased heart rate, perspiration and the redistribution of blood from the digestive system to the muscles. As they have opposite effects, the sympathetic and parasympathetic systems are said to be **antagonistic**.

Nerves carry electrical impulses from one part of the body to another. Each nerve is a bundle of nerve fibres, and is made up of a large number of nerve cells or **neurones** (see Figure 13.4). Neurones consist of a nerve cell body and nerve fibres. The **cell body** has the nucleus and cytoplasm which contains organelles, including many mitochondria to generate the large amounts of energy needed. The fibres are **dendrites** which carry impulses towards the cell body and a single **axon** which carries impulses away from the cell body. Long fibres are covered in a **myelin sheath** which greatly increases the speed of transmission of the impulse. Myelination is not complete at birth and so nervous control increases over the first two years as more neurones become fully myelinated.

Like all cells, neurones have variations in structure which adapt to their function.

- **Motor neurones** – have short dendrites, in contact with other neurones in the CNS, a cell body and a long myelinated axon that carries impulses to many connections on a muscle.
- **Sensory neurones** – have dendrites in contact with sense receptors. These merge to form a myelinated fibre which carries impulses to the cell body, a short axon and connections with neurones in the CNS.
- **Associative neurones** – found only in the CNS where their many dendrites form contacts with one another. These form the complex network of interconnections which allows the system to carry out its incredible functions, and to learn, remember and adapt in the way it does.

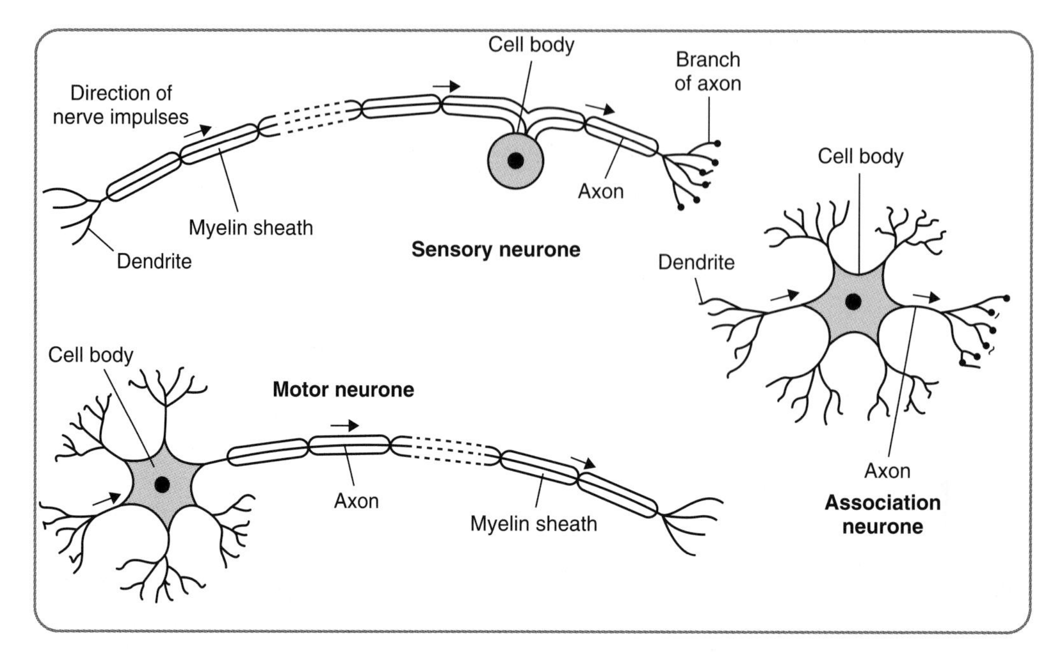

Figure 13.4 Structure of neurones

For Practice

1 Redraw and complete the summary of the important parts of the nervous system shown in Figure 13.5.

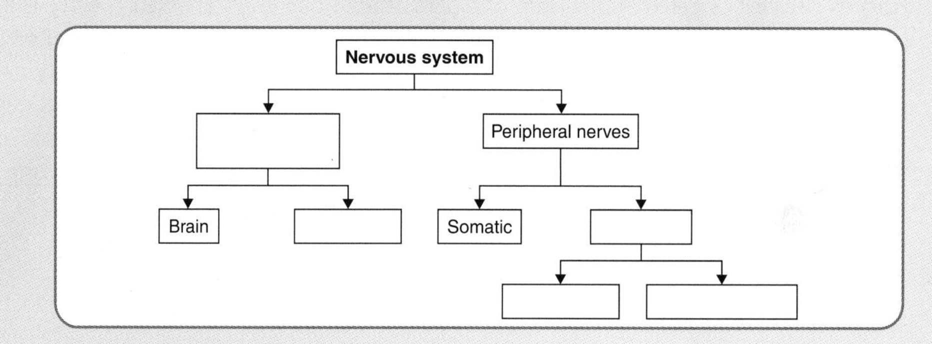

Figure 13.5 The nervous system

2 Name the type of neurone which:

(a) carries impulses from the sense organs to the CNS

(b) carries impulses from the CNS to muscles

(c) is found only in the CNS.

3 Rearrange the main parts of a neurone in the order in which an impulse passes through them: cell body, axon, dendrites.

For Practice *continued* ➢

For Practice continued

4 Why is it necessary for a neurone to have a large number of mitochondria?

5 Copy or trace the diagrams in Figure 13.6 and add the correct labels. For each label add a brief note describing the function of the structure.

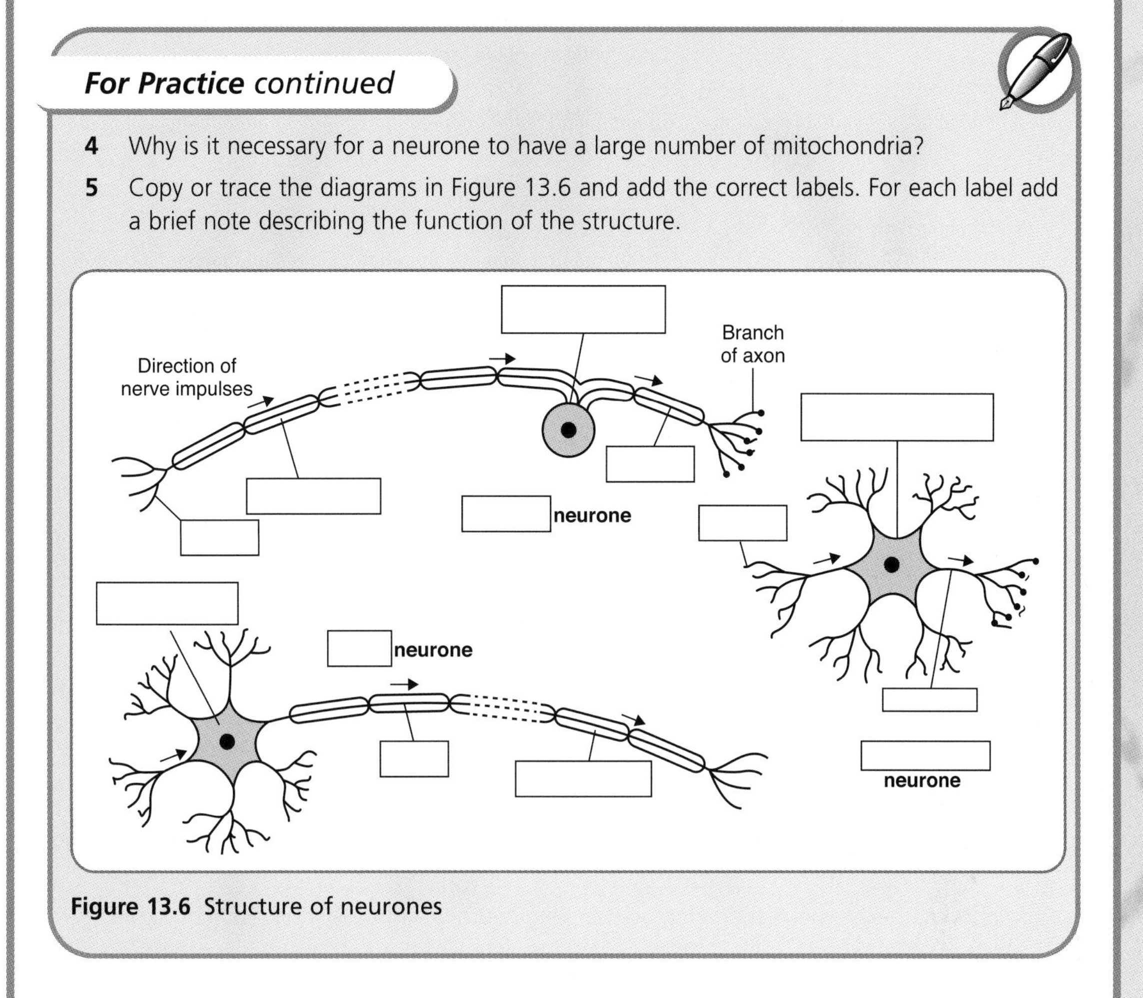

Figure 13.6 Structure of neurones

Synapses

The key to the incredible abilities of the nervous system is in the adaptability of the connections amongst neurones. This is the result of the functioning of **synapses**, which are the tiny gaps between neurones. Impulses cannot cross the synapse. The structure of a synapse is shown in Figure 13.7.

The impulse is created in the next cell when enough of a chemical diffuses across the gap to stimulate special receptors in the membranes of the following cells. These chemicals are called **neurotransmitters**, and important examples include **acetylcholine** and **noradrenaline**. If sufficient transmitter molecules reach the membrane of the next neurone an impulse is triggered in that cell, and so on. Weak stimuli are filtered out because not enough transmitter molecules reach the next neurone.

The transmitter molecules have to be removed between impulses so that the signal does not go on for ever. Acetylcholine is broken down by an enzyme; whilst noradrenaline is reabsorbed into the pre-synaptic cell. It is the receptors that control the transmission of the impulse. The same transmitter molecule can either excite or inhibit receptors.

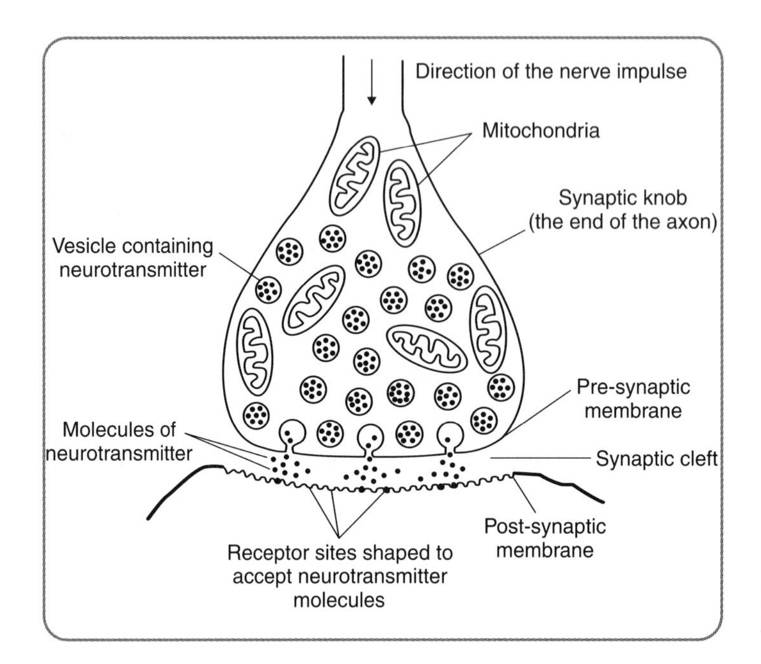

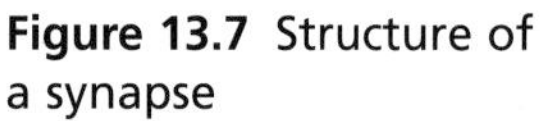
Figure 13.7 Structure of a synapse

Converging and diverging neural pathways

Converging pathways are where several neurones feed impulses to one neurone. For example, the light-sensing rod cells in the eye are connected several at a time to each sensory neurone, allowing even a very weak stimulus from a dim light to be detected in the brain.

Diverging pathways mean that an impulse in one neurone can have a simultaneous effect in many parts of the body. For example, a diverging neural pathway from the hypothalamus can stimulate many sweat glands and arterioles simultaneously. In the same way, a single stimulus into a diverging pathway can operate many different muscles, allowing complex movements such as delicate manipulation, or balancing.

It is possible to suppress reflexes or sensory impulses. For example, a bad taste may result in the reflex of being sick, but the relevant muscles may receive a stronger stimulus from another part of the brain which will hopefully make you resist if you are at a posh banquet. This ability is called plasticity of response.

For Practice

1 (a) What are neurotransmitters?

(b) How do neurotransmitters cross synapses?

(c) Give two examples of neurotransmitters. For each say how it is removed from the synapse.

2 (a) What are converging neural pathways and what is their main function?

(b) What are diverging neural pathways and what is their main function?

3 What is meant by 'plasticity of response'?

4 Copy or trace the synapse diagram in Figure 13.8 and add the correct labels. For each label add a brief note describing the function of the structure.

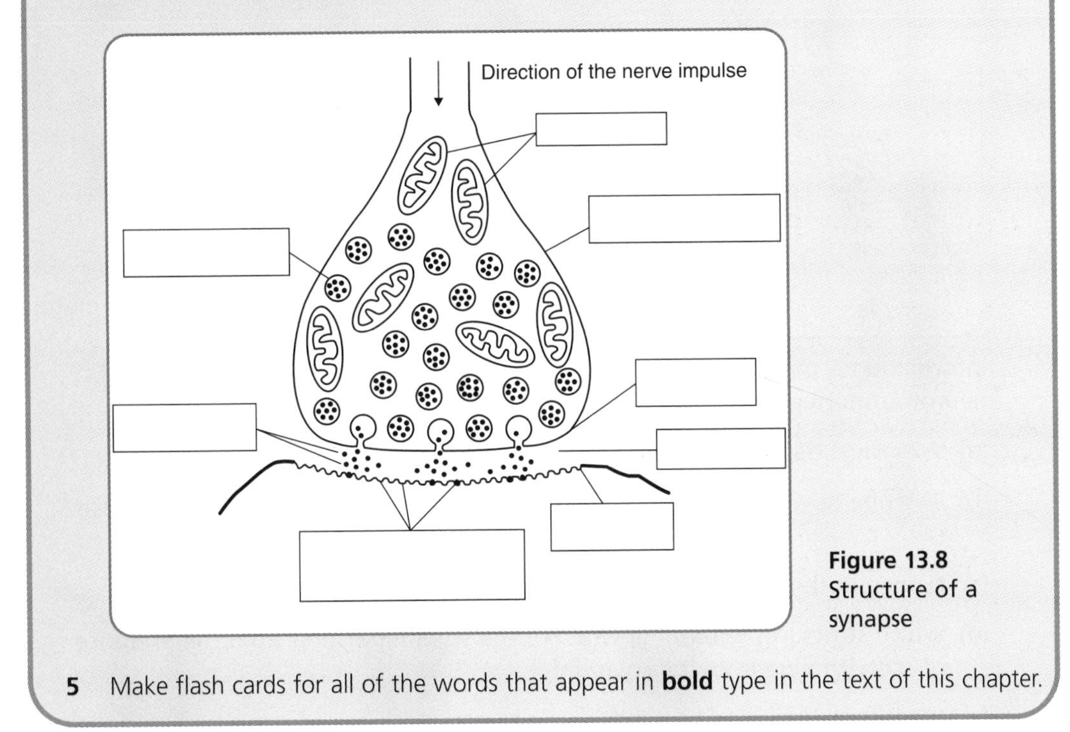

Figure 13.8 Structure of a synapse

5 Make flash cards for all of the words that appear in **bold** type in the text of this chapter.

Exam Questions

1 Figure 13.9 shows divisions of the nervous system.

(a) State the function of the somatic nervous system. *(1)*

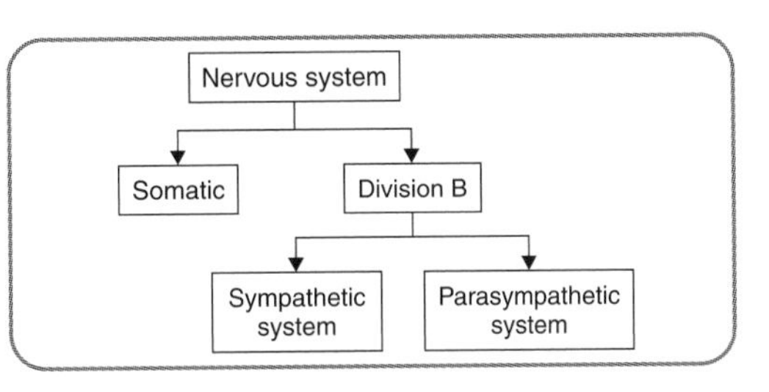

Figure 13.9 Divisions of the nervous system

Exam Questions continued

(b) Name division B of the nervous system. *(1)*

(c) The sympathetic and parasympathetic systems are antagonistic.

(i) Describe the antagonistic effect of the sympathetic and parasympathetic systems on the heart. *(1)*

(ii) Describe the effect of parasympathetic stimulation on smooth muscle in the wall of the small intestine. *(1)*

2 Figure 13.10 shows a motor neurone.

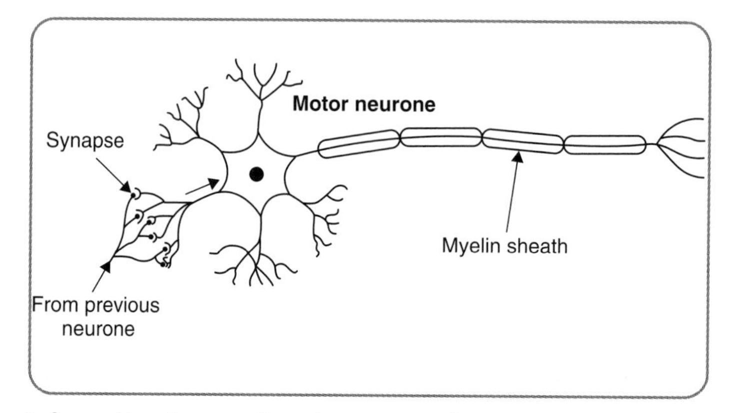

Figure 13.10 A motor neurone

Information is passed at the synapse from neurone to neurone by neurotransmitters.

(a) Why must neurotransmitters released at the synapse be removed rapidly? *(1)*

(b) Describe how the neurotransmitter acetylcholine is removed at the synapse. *(1)*

(c) Name another neurotransmitter. *(1)*

(d) Which structures determine whether the stimulus arising from the presence of acetylcholine is excitatory or inhibitory? *(1)*

(e) What is the effect of myelination on the transmission of impulses along the axon? *(1)*

3 Give an account of the transmission of nerve impulses under the following headings:

(a) The structure of neurones *(4)*

(b) The synapse. *(6)*

Answers

1 (a) Controls conscious or voluntary activity.

(b) Autonomic.

(c) (i) Sympathetic increases heart rate and parasympathetic decreases heart rate. Both for 1 mark. Note that you need to give both sides of the answer when asked for comparisons – you cannot simply assume that the other half of the answer is obvious.

(ii) Speeds up contractions or peristalsis.

2 (a) It is because otherwise it would continue to have its effect so the synapse could not function.

(b) It is broken down by an enzyme.

(c) Noradrenaline.

(d) Receptors.

(e) You should know that it speeds it up.

3 (a) The structure of neurones

(b) The synapse.

Nos	Mark scoring points		Comments
	(a) The structure of neurones		*Write the sub-heading*
1	Nerve cell body with nucleus	1	*You simply need to describe a neurone, including each structure. You could make a labelled drawing, but only the labels would score marks, so make sure you label everything*
2	Dendrites	1	
3	Axon	1	
4	Myelin sheath	1	
5	Direction of impulse (from dendrites through cell body to axon)	1	
6	Distinction between sensory, motor, intermediate neurones	1	
			*Any four of these points would give you the **maximum of 4** for this part of the answer, but obviously you should include as many as possible – some might not count*

Answers continued

Nos	Mark scoring points		Comments
	(b) The synapse		
7	Synapse is the junction between neurones	1	*It is always a good idea to give a definition*
8	Tiny gap or no connection between cells	1	*The definition is followed by a clear but simple description of how a synapse works. Notice that marks are given for each detail, so make sure that you include every stage – don't fall into the trap of thinking that a point is too simple to mention*
9	Use of terms: neurotransmitter, acetylcholine, noradrenaline	1	
10	Transmitters released from vesicles following arrival of impulse	1	
11	Diffuse across gap	1	
12	Bind to receptor	1	
13	Impulse in next neurone stimulated	1	
14	Impulse can be excitatory or inhibitory	1	*And finally some general information. The order is not important, but it always helps to put things in an organised way.*
			Any six of these points would be enough to get the ***maximum 6 marks*** *for this part*
15	Transmitter must be inactivated or removed	1	
16	By absorption or broken down by enzymes	1	
17	Many mitochondria – high energy need	1	

Chapter 14

MEMORY

Memory is one of the most important abilities. Without it there could be no learning, no experience, no possibility of solving problems, no recognition, and, arguably, very little in the way of personality.

Localisation of memory in the brain

Right in the core of the brain, and closely connected to other vital structures such as the pituitary, which is largely responsible for hormonal activity, and the frontal lobes which deal with conscious thought and reasoning, the **limbic system** is thought to be the part of the brain chiefly involved with memory.

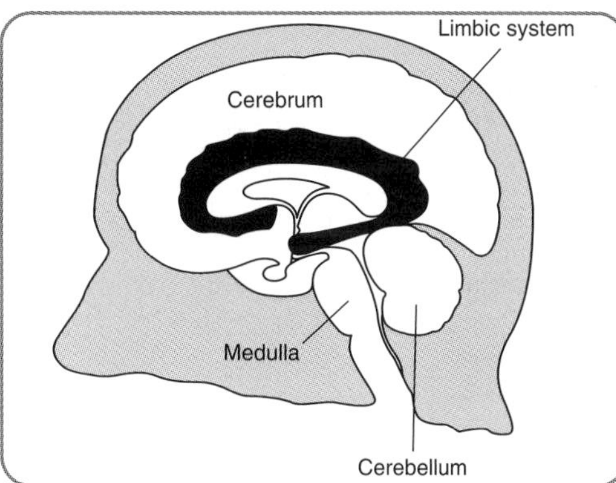

Figure 14.1 Limbic system

Encoding, storage and retrieval

It is convenient to think of the way memory works in terms of these categories. **Encoding** is the process of selecting what to remember from the millions of sensory impulses received by our brain every second, and then converting the result into a form that can be stored. Various types of encoding are used, including converting sensory events into pictures, sounds or words. **Storage** involves keeping the memory in our brain, and available for use, for some time. Depending on the circumstances and the efficiency of encoding, this may be anything from seconds to a lifetime. As with encoding, storage seems to happen in many different ways for different purposes. The final stage is **retrieval**. The test of success is our ability to gain access to the stored information and the degree of accuracy with which we can do so. When we forget something, the failure could be in any of these stages.

Short-term memory

An important part in the process of filtering out unimportant information and selecting what will be remembered takes place in the first few seconds. Short-term memory has a very limited capacity. Try repeating a list of random numbers or words. You will find that the limit is somewhere between five and nine bits of information. This is called the **memory span**. If any more are added, then some of the earlier ones are displaced to 'make room' for the new ones. In addition, short-term memory can retain information for no more than a few seconds – half a minute would be exceptional.

The usefulness and capacity of short-term memory can be greatly increased by organising the material. For example, the letters W E W O N T W O O N E H E C T … would completely overflow the short-term memory, but in the form of meaningful words 'We won 2-1, Hector scored both goals' would easily fit into short-term memory as seven or so pieces of information. Grouping several smaller units into a single meaningful piece of information is called **chunking**. Other examples of chunking might include encoding a description into the form of a single image, or comparing the detail of a new event with a previous experience that is already in memory.

For Practice

1 Get somebody to read out a list of random numbers or letters and investigate your own memory span. It is likely to be somewhere between five and nine items.

2 Copy and complete the table below to show detail of encoding, storage, retrieval and chunking.

Process	Detail
Encoding	

Transfer of information from short-term memory to long-term memory

Short-term memory stores information for only a few seconds. Information that is to be remembered for longer periods of time must be transferred to long-term memory. **Long-term memory** seems to have an unlimited storage capacity, and is often capable of retaining information for the entire lifetime of an individual. In order to make use of it successfully, however, the information must be efficiently transferred from short-term memory. Various processes can assist the transfer, the most important of which are:

- rehearsal
- organisation
- elaboration of meaning.

Hints and Tips

Long-term memory is the key to exam success, so it is certain that understanding rehearsal, organisation and elaboration of meaning are well worth learning about, and incorporating into your study skills.

- **Rehearsal** is when the information is repeated many times, making it more likely to be transferred into long-term memory. It can involve re-reading short sections of written material, or copying it in your own words – in fact any of a wide range of repetition strategies. Research suggests that the success of rehearsal is increased by the frequency of repetition, and by short intervals between repetitions. Rehearsal is the basis of rote learning where information, for example a multiplication table or the words of a song, are repeated so often that they are never forgotten.

- **Organisation** involves arranging information so that it is grouped or categorised in a way that seems logical to the learner. Long-term memory stores information in a variety of groups and sections, just as the shelves and drawers in your bedroom allow you to store items in an organised way with areas allocated to shirts, trousers, underwear, socks and so on. Organised information is much more effectively transferred to long-term memory than random pieces of information.
- **Elaboration** of meaning is the process of making facts easier to transfer and store by building them into a bigger story. A name, such as Angus or Morag, may or may not be recalled, but if the name is associated with a range of other information, such as a mental picture of the person, some information about habits and personality, and other memories of behaviour, events and possibly even fragrances, then all the details of the 'package' are certain to be more effectively stored and recalled. In the same way, facts that need to be learned for examinations are much more securely encoded into long-term memory if they are built into a wider structure of knowledge and understanding of a particular area of study.

For Practice

Briefly note down some specific examples of ways in which people use each of the methods of rehearsal, organisation and elaboration of meaning in order to make their memory and learning more effective.

The serial position effect

When information is viewed in a sequence the first and last few pieces of information are remembered best. This is called the **serial position effect**. The first few are remembered because there are not yet enough items to prevent some rehearsal before the later ones crowd in. The last few have not yet been displaced from the small storage capacity of the short-term memory.

Figure 14.2 shows the results of an experiment in which subjects were shown 20 familiar objects. Teachers, authors and film makers constantly make use of this effect by starting and ending with what they want to be most memorable, and advertisers pay considerably more for the first and last slots in commercial breaks.

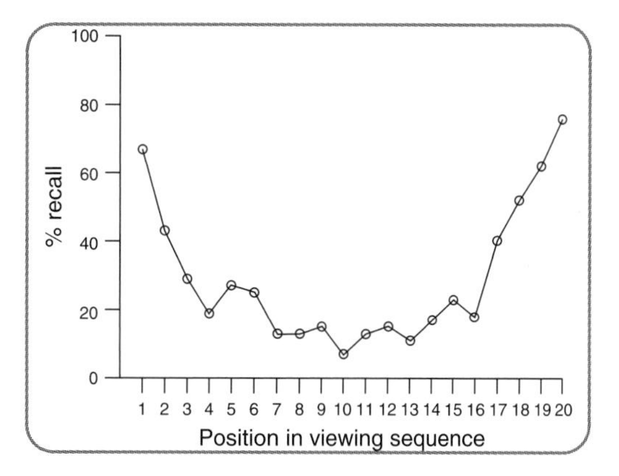

Figure 14.2 Experiment results

Contextual cues

Retrieval is almost always helped by recall or re-experience of the situation surrounding the moment when we first encoded a memory. For example, a particular smell or piece of music can bring a whole series of memories back to mind. Such events are called **contextual cues**. Often the context will influence the organisation of the memory and it is common for the same memory to be associated with a variety of quite different contextual cues, any of which is capable of promoting the retrieval of events from long-term memory.

For Practice

Play a piece of music you loved two years ago. As you listen, memories of the time are sure to flood back.

What other senses can trigger memories? Which sense is most powerful in this respect for you? Research suggests that for many people, the sense of smell can be the most evocative.

The evidence for a molecular basis for memory

Alzheimer's disease involves progressive loss of memory as well as many other distressing symptoms. Researchers have investigated changes in the limbic systems of sufferers. Two chemicals in particular have been associated with the memory process. Alzheimer's patients are known to lose cells in the limbic system which produce **acetylcholine**, and this is possible evidence that acetylcholine is involved in linking neurones to form memories. Similarly the limbic system is particularly rich in receptors for **NMDA** (a simple protein-like molecule) and this suggests that NMDA has a role in memory storage.

For Practice

Make flash cards for all of the words that appear in **bold** type in the text of this chapter.

Exam Questions

1 The conversion of information into a form which the memory can accept is:

A rehearsal

B chunking

C encoding

D organisation.

2 Which statement correctly describes a chemical deficiency in Alzheimer's disease? Loss of:

A acetylcholine produced in the limbic system

B acetylcholine produced in the cerebellum

C NMDA produced in the limbic system

D NMDA produced in the cerebellum.

Exam Questions continued

3 The table shows the procedures used in an investigation into short-term and long-term memory.

Group	Pace of reading a list of 20 words	Time of testing recall
Group 1	1 second between words	Immediate
Group 2	1 second between words	30 seconds' delay
Group 3	5 seconds between words	Immediate

In Figure 14.3, Graph A shows the results from groups 1 and 2.

Graph B shows the results from groups 1 and 3.

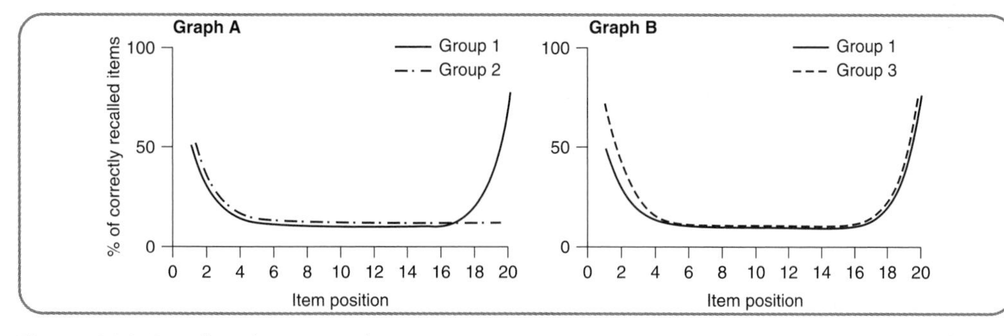

Figure 14.3 Results: short-term/long-term memory

(a) What term is used to describe the pattern of results for group 1? *(1)*

(b) From Graph A, what evidence supports the statement that delay before recall has an effect on short-term memory? *(1)*

(c) From Graph B, describe the effect of spacing on long-term and short-term memories. *(1)*

(d) Rehearsal is one technique used to improve transfer of memory from short-term to long-term memory. Describe one other technique used in memory transfer. *(2)*

(e) What is meant by 'memory span' when applied to short-term memory? *(1)*

(f) What is meant by 'chunking'? Give an example. *(2)*

4 Give an account of memory under the following headings:

(a) Encoding *(6)*

(b) Storage *(2)*

(c) Retrieval. *(2)*

Answers

1 **Answer** C

2 **Answer** A

3 (a) Serial position effect.

(b) Group 2 forget the last items.

(c) Longer spacing increases the recall for earlier items.

(d) Chunking is grouping items so that each group works as a single item.
or
Organisation is the classification of items into groups with a similar characteristic.
or
Elaboration is adding a framework of supporting information about the item.

(e) This is the limited number of items that can be stored at one time.

(f) Grouping pieces of information into a meaningful unit so that they can be stored in short-term memory. Remembering words as meaningful concepts rather than individual letters.

4

Nos	Mark scoring points		Comments
	(a) encoding		*Write the sub-heading*
1	The way information is entered into the memory	1	
2	Examples from visual, sound, taste, smell, tactile and semantic	2	*4 examples for 2 marks; 2 or 3 examples for 1 mark*
3	Enhanced by rehearsal, organisation and elaboration	1	
4	Description or example of rehearsal	1	
5	Description or example of organisation	1	
6	Description or example of elaboration	1	
7	Description or example of mnemonics	1	
			Any of the above for a ***maximum of 6 marks***
	(b) storage		*Write the sub-heading*
8	STM has a limited capacity of (about) 7 items	1	

Answers continued

Nos	Mark scoring points		Comments
9	More information can be stored by chunking	1	
10	STM information transferred to LTM *or* lost *or* displaced	1	*Only one fate of STM items required for the mark*
11	LTM has unlimited capacity	1	
12	Memory in limbic system *or* hippocampus	1	Either site of memory gains the mark
			Any of the above for a ***maximum of 2 marks***
	(c) retrieval		*Write the sub-heading*
13	Getting information out of memory	1	
14	Aided by contextual cues	1	
15	An example of a contextual cue from sight *or* smell *or* sound	1	*Only one example of a contextual cue required for the mark*
16	Description of serial position effect	1	
			Any of the above for a ***maximum of 2 marks***
			Maximum = 10 marks

Chapter 15

BEHAVIOUR

Factors influencing the development of behaviour

All aspects of development, including behaviour, are influenced by **inheritance**, **maturation** and the **environment**. It is impossible to identify behaviours that are influenced by only one of these. Behaviours are dependent on different combinations of these factors and the extent to which each is involved is one of the main interests in the study of behaviour.

Maturation

Maturation is an inherited sequence of developmental stages. That the sequence is inherited is evidenced by the fact that it is similar in children of all cultures and environments. The rate of development, however, can be enhanced or limited by genetic or environmental factors.

The stages of development of walking are a good example of development that is heavily influenced by maturation. All children pass through the stages shown in Figure 15.1. The sequence is universal, although the timing of each stage can vary considerably.

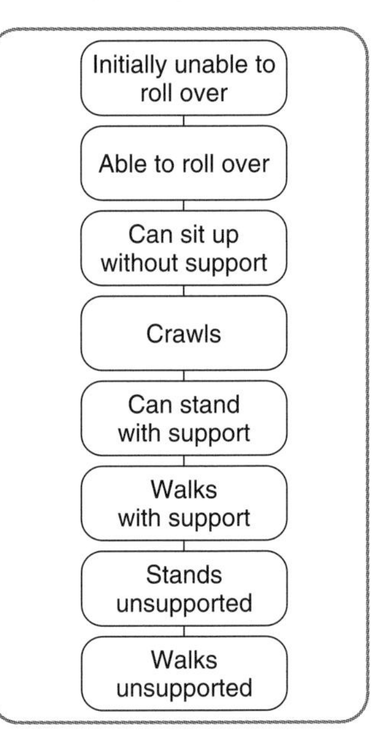

Figure 15.1 Stages in the development of walking

Although some of the stages are limited by the development of the required muscle and bone structure, for example to support the weight of the body, much of the maturation process depends on the development of the nervous system. Infants have neurones which are only partially myelinated. Myelin is the fatty covering which surrounds the axon and makes the neurone capable of transmitting impulses much faster. As infants develop, **myelination** is completed and motor and sensory impulses become fast enough to allow the development of increasingly complex behaviours.

Research with identical twins has shown that training, practice and encouragement cannot greatly accelerate the age of walking. The development of the behaviour is limited by the rate of maturation.

Inheritance

Genes inherited from parents clearly have an important influence on the potential for the development of behaviour. The sequence of developmental stages depends on the development of a fully functional nervous system that is working in an average kind of way; but any problems can limit potential development.

Some inherited conditions affect the development of the nervous system, and so can affect behaviour.

Phenylketonuria (PKU) is caused by a defect in a single gene which makes the sufferer unable to produce an important enzyme to process the amino acid phenylalanine. This results in the build-up of toxins which disrupt brain cell metabolism and prevent proper brain development. Untreated children have severely limited mental development and die in early adulthood. Happily, post-natal screening and a phenylalanine-free diet ensures normal brain development.

Huntington's chorea depends on a single dominant gene and so it may be inherited from either parent. The condition shows no symptoms until sufferers are in their late thirties. From this time on there is progressive mental deterioration followed by paralysis, dementia and death. Sadly there is no known cure.

Environment

It is difficult to measure how much influence environment has on behavioural development – we will return to the subject in the section 'Communication and social behaviour' on p. 150. Attempts to accelerate behavioural development, for example efforts to train infants to walk early, have very limited success because the process is limited by processes of maturation, such as myelination, bone and muscle development. The same is true for other skills such as talking and toilet training.

One way of illustrating the influence of environment is by studying **identical twins**. Since these individuals have identical genes, it is assumed that genetic and maturation factors are the same, and that behavioural differences are due to the environment. Such studies suggest that environmental effects are significant, but limited.

For Practice

1 Construct a table like the one below and add examples of behaviour which are significantly influenced by each other.

Maturation	Inheritance	Environment

The inter-relationship between maturation, inheritance and the environment

No aspect of behaviour attracts more heated controversy than **intelligence**. It is not long since attempts were made to assert that it is controlled by genes alone, and that maturation and environment had no effect. Indeed there are still remnants of such primitive notions to be found in less informed parts of the educational system. The (completely mistaken) idea that an individual is born with a certain intelligence, and that this will not change through life, no matter what training or learning is involved, still seems to appeal to both teachers and underachieving pupils alike, as a convenient excuse.

In fact much of the problem lies in the fact that intelligence is a very poorly defined term, with the result that it is used to mean different things to different people. In modern scientific usage intelligence is regarded as the combination of a wide range of intellectual skills. These would include, for example, musical, mathematical, artistic, physical, and other abilities.

So-called IQ tests focus on only a small group of these skills, and make little or no allowance for maturation or environment.

The inherited component of intelligence is extremely difficult to investigate. Firstly, it is **polygenic**, which means that it depends on a large number of different genes, each of which will be subject to a range of natural genetic variation. This produces a vast range of possible combinations for any two parents. In addition, the effects of environment and maturation are uncontrolled and almost impossible to separate.

Communication and social behaviour

The behaviours selected by an individual are the result of events in the nervous system. The sense organs detect stimuli from the surroundings and transmit electrical impulses along the sensory neurones to the CNS. Different parts of the brain analyse the many pieces of information in the light of past experience. Usually it is decided that nothing needs to be done; but sometimes impulses are sent though motor neurones to muscles to make a response.

The effect of infant attachment

Human children are dependent on adults for a long time. During the early part of this period, critical stages of development must take place to allow the later development of important social skills such as communication. The long period of **dependency** also provides time and opportunities for learning with some protection from the dangers of the error part of 'trial and error'.

The effect of communication

Human **language** is, by far, the most complex and sophisticated communication vehicle in the animal world. The ability of humans to communicate through language makes much of their behaviour unique. Words can symbolise complex ideas and concepts, and sentences build them into packages of detailed information. This makes language, whether written or spoken, an incredibly fast and effective means of communicating our needs, emotions and views to other members of society. Much of human behaviour is unique to our species because of the ability to communicate through language.

Language, however, is not the only means of communication. **Non-verbal communication** starts even before the birth of a baby, and continues to be important throughout life.

Before they develop language, infants rely solely on non-verbal communication, and it is therefore vital in establishing the adult–child bonding that is so important for the successful development of the child. For example, smiling is a particularly effective way of reinforcing bonds between individuals of all ages.

Non-verbal communication also enhances the use of language, by conveying additional information about mood, emotions and attitudes. Some of this is subconscious and so we are often unaware of the information we are conveying.

It may even be the case that non-verbal signals contradict the spoken words, and in such cases the message of the non-verbal signals, or body language, may be more influential than the words.

For Practice

1 What do these gestures mean?

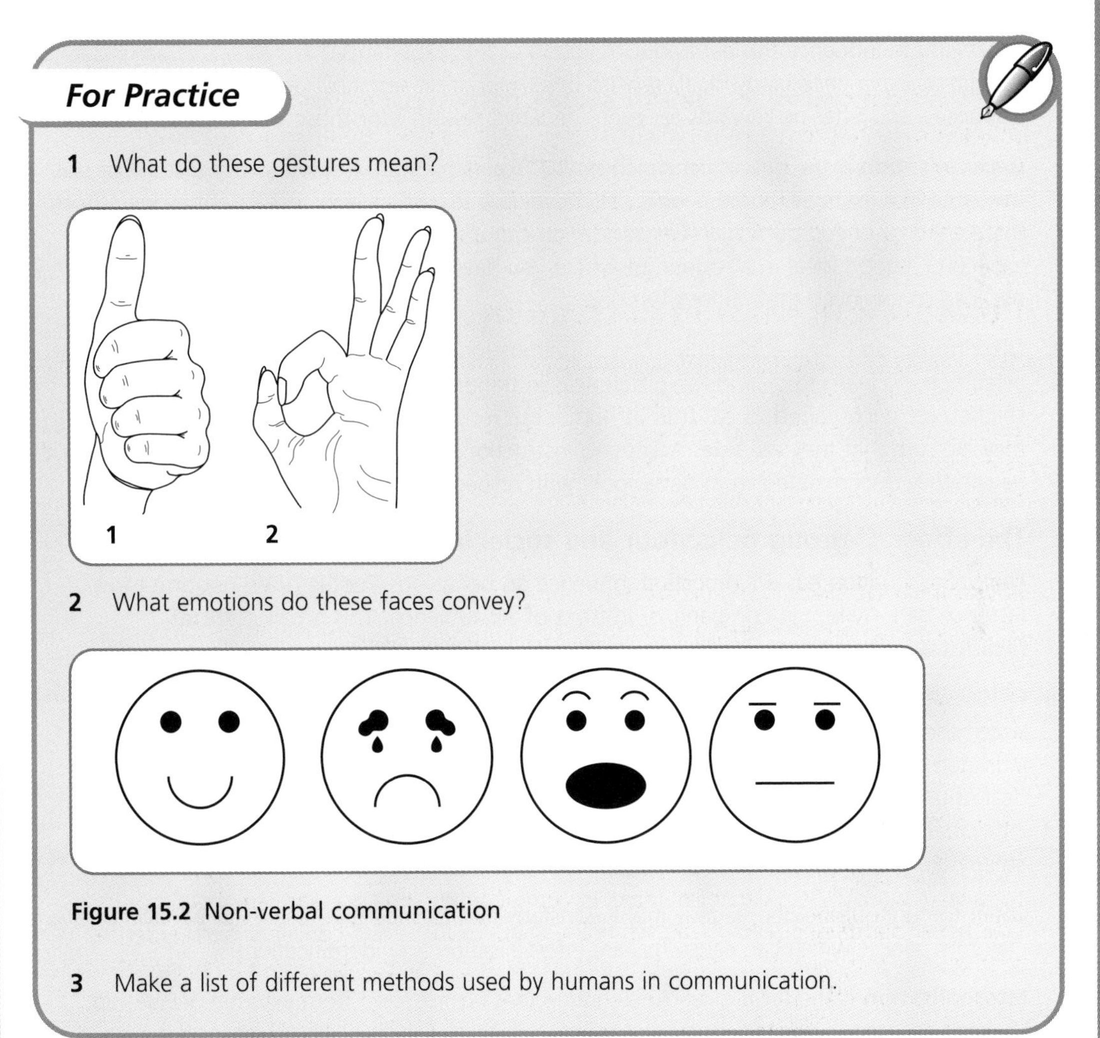

2 What emotions do these faces convey?

Figure 15.2 Non-verbal communication

3 Make a list of different methods used by humans in communication.

The effect of experience

Behaviour can be affected by experience in a variety of ways.

Practice is the repeated use of a skill. In the case of motor skills it results in a motor pathway in the brain being established. This **motor memory** can be seen, for example, in the skills of riding a bike or driving a car where, eventually, the skill becomes almost automatic.

Imitation, the copying of observed behaviours, is the method by which much of human behaviour is learned. Training people in almost any skill uses this method to a considerable extent.

Trial and error learning occurs because individuals are likely to repeat successful behaviours, which give them some benefit, whilst they are likely to avoid repeating behaviour that offers no benefit. Behaviours that bring a reward are **reinforced**, whilst those that bring no reward, or even a negative outcome, are likely to disappear (**extinction**).

Cultural transmission is the passing of information from generation to generation, and is an important factor in the evolution of behaviour. **Shaping** is the name given to the process of seeking to influence the behaviour of young or less experienced individuals in particular directions. Experienced individuals reward behaviours that resemble desired or appropriate responses, and offer no rewards (or even 'negative reward') for those that are not so good.

Generalisation is the process in which similar experiences are treated as being more or less the same. For example, some people may develop a fear of all dogs because they were once threatened by one in particular. Generalisation can also relate to pleasant experiences like expecting a good meal in a restaurant where you have eaten well before, or buying a book because you enjoyed the author's last one.

Discrimination is an equally important but almost opposite skill to generalisation as people often need to learn to distinguish differences in things that are quite similar. For example, children are encouraged to eat fruit at home, but must learn not to eat wild berries until they are sure that they are safe. Accepting instructions or treats from friends and family must be carefully discriminated from behaviours with respect to strangers.

The effect of group behaviour and social influence

The social situation has an important influence on behaviour. People often perform tasks faster or better when in company, or in front of an audience. This is called **social facilitation**.

On the other hand, people sometimes take part in anti-social or inappropriate behaviour which they would not think of doing when alone. They seem to lose their feeling of individual identity and take on a group identity. This is known as **deindividuation**. A great deal of unacceptable and even inhumane behaviour can be traced to deindividuation. Every individual has a responsibility to retain his or her individual judgement in matters of acceptable and unacceptable behaviour.

Often throughout life people may make decisions that change their beliefs or behaviour. Two major influences which can cause this are internalisation and identification.

Internalisation is the term given to a change that has resulted from external persuasion. We may become aware of a body of evidence which convinces us that a change is a good idea and so make the change part of our behaviour. Examples of internalisation might include deciding to stop smoking, to adopt a healthier lifestyle, or for a criminal to 'go straight'.

Identification is when we admire or respect someone, and make adjustments to our behaviour to become more like the person who influences us. We can be influenced in this way by friends or colleagues, or by teachers, sporting heroes and celebrities.

Since much of the world depends on influencing our behaviour in terms of buying things or accepting beliefs such as politics and religion, we are constantly bombarded by all sorts of influences. Advertising, education, fashion and much else seeks to shape our behaviour, and in almost all of the materials they produce they rely heavily on the tools of internalisation and identification.

For Practice

1. Each of the categories of experience which influence behaviour (**bold** type) have shaped your own behaviour and that of others. Make a table with the categories in one column and add specific examples of how your own behaviour has been modified in that way.
2. Consider how resistant to deindividuation you are. Have you ever behaved in a way you are not proud of because of it – probably best not to write down your answers to this!
3. Are there role models with whom you identify sufficiently to modify aspects of your behaviour?
4. Make flash cards for all of the words that appear in **bold** type in the text of this chapter.

Exam Questions

1 The list below shows stages in the development leading to walking in infants.

1 Stands with support

2 Sits without support

3 Stands alone well without support

4 Walks with support

The correct sequence in development leading to walking is shown by:

A 2 – 1 – 3 – 4

B 1 – 2 – 4 – 3

C 2 – 1 – 4 – 3

D 1 – 2 – 3 – 4.

2 A child, frightened by a German Shepherd dog, still feels safe with the family collie. This behaviour is an example of:

A generalisation

B deindividuation

C internalisation

D discrimination.

Exam Questions continued

3 An individual in a group will often demonstrate anti-social behaviour which they would not carry out if they were on their own. The behaviour of the individual within the group is called:

A imitation

B deindividuation

C social facilitation

D internalisation.

4 Figure 15.3 shows the time taken by a pupil to complete a maze over a series of trials and the number of errors in each of the trials.

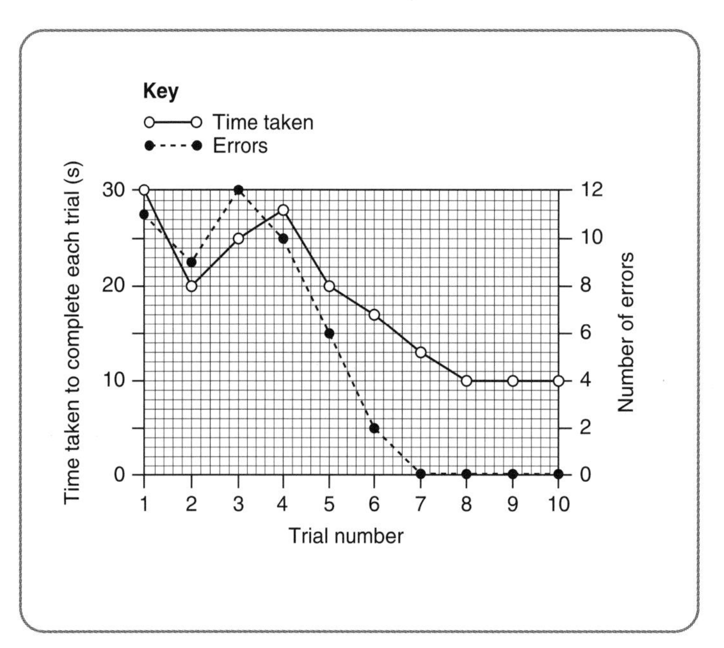

Figure 15.3 Maze trial results

Which of the following statements is correct?

A When one of the trials took 20 seconds the number of errors was 6.

B The number of errors was greatest in the first trial.

C The shortest times taken were the only ones free of error.

D The shortest time to complete the maze correctly was 4 seconds.

5 Describe behaviour under the following headings:

(a) Shaping and rewarding *(3)*

(b) Deindividuation *(3)*

(c) Influences that change beliefs. *(4)*

Answers

1 C

2 D

3 B

4 A

5

Nos	Mark scoring points		Comments
	(a) Shaping and rewarding		*Write the sub-heading*
1	Shaping is when the parent or trainer obtains a desired response in the child or learner developing a skill	1	
2	Desired behaviour pattern is rewarded	1	
3	Desired response is then managed by child or learner unaided	1	
4	Rewarding is when parents or trainer reinforce desired behaviour with praise, treats, etc.	1	
			*Any of the above for a **maximum of 3 marks***
	(b) Deindividuation		*Write the sub-heading*
5	Individual loses personal identity when in the group	1	
6	Behaviour becomes less and less acceptable compared to norm	1	
7	Individual in the group behaves in a way they would not on their own	1	
8	Example to demonstrate deindividuation	1	
			*Any of the above for a **maximum of 3 marks***
	(c) Influences that change beliefs		*Write the sub-heading*
9	Internalisation	1	
10	Individual's beliefs or behaviour changed as a result of persuasion	1	
11	Example to demonstrate internalisation	1	
12	Identification	1	

Answers continued

Nos	Mark scoring points		Comments
13	Change beliefs to be more like someone that they admire	1	
14	Example to demonstrate identification	1	
			Any of the above for a **maximum of 4 marks**
			Maximum = 10 marks

POPULATION GROWTH AND THE ENVIRONMENT

Human population growth

Every species produces more offspring than are needed to replace the adults when they die. As the number that can live in a particular area – the **carrying capacity** – stays much the same from year to year, most of the offspring must die. We have probably all heard versions of the scary numbers which result if, for example, two mice are left to breed. Within five years there would be so many that they would weigh more than the entire planet Earth. And this is true, except that lack of food, habitats, water, the effects of predators, parasites, disease, and many other factors keep the numbers within limits.

The human species is regulated by the same factors; but by a combination of thought and aggression we have reduced the effect of predators and disease and have hugely expanded the carrying capacity of the environment at the expense of other species.

History of human population growth

Archaeological evidence suggests that modern humans reached a population of four or five million individuals and remained more or less at this level until somewhere around 10 000 years ago. To give this some perspective, this means that the total human population of the entire world was the same as now live in the central belt of Scotland between Edinburgh and Glasgow.

At this time, the development of agriculture, and the permanent settlement which this allowed, led to improvements in social care and co-operation, and increased development of culture and learning about health, protection and nutrition. The result was an increase in the carrying capacity of the environment, and a steady rise in population. This growth appears to have been more or less linear until as recently as a few hundred years ago. Between 8000 BC and 1600 AD the population rose to somewhere between 300 and 500 million.

Since then population growth has moved from a linear to an **exponential increase**, where the *rate* of increase gets faster and faster as the numbers increase. As a result the numbers have increased at a terrifying rate. Figure 16.1 shows just how startling this is.

The significant changes shown in Figure 16.1 can be explained relatively simply. Before agriculture was developed, humans were **hunter-gatherers**, leading a nomadic life and finding food and shelter as well as possible. Bad years, drought, fires or floods would have a catastrophic effect, and there was no way of storing food or water for extended times of shortage. The climate would strictly limit the extent to which the population could expand, and so the carrying capacity would remain fairly stable, with the population probably nowhere near reaching that level.

Agriculture brought with it increased yields, permanent settlements, and the ability to store resources as insurance against bad times. Irrigation and the use of composts and manure

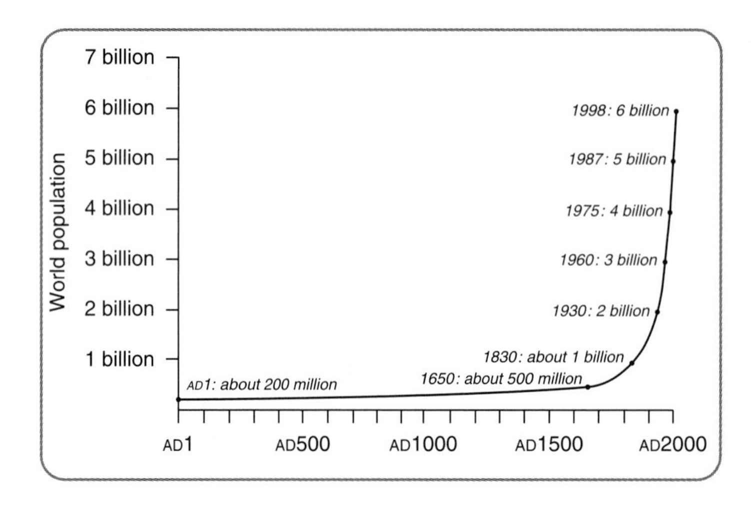

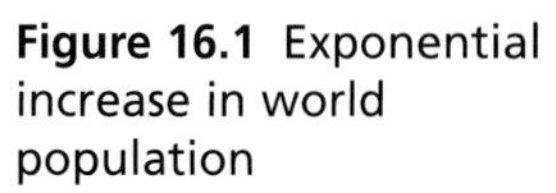
Figure 16.1 Exponential increase in world population

extended the area of land that could sustain human life, and the carrying capacity increased considerably.

By the end of the seventeenth century, industry and science began to generate a further population explosion – the exponential phase which is currently taking place, and which cannot be sustained for much longer. There are fortunately some tentative signs of a slowing of growth in the last twenty years.

Demographic trends show what is happening to the populations of particular countries or groups of countries. The death rate has fallen; this has happened in both developing and developed countries. In the developed world the birth rate has also fallen, leading to a slowing of the population increase, whilst in developing countries the birth rate remains very high (see Figure 16.2).

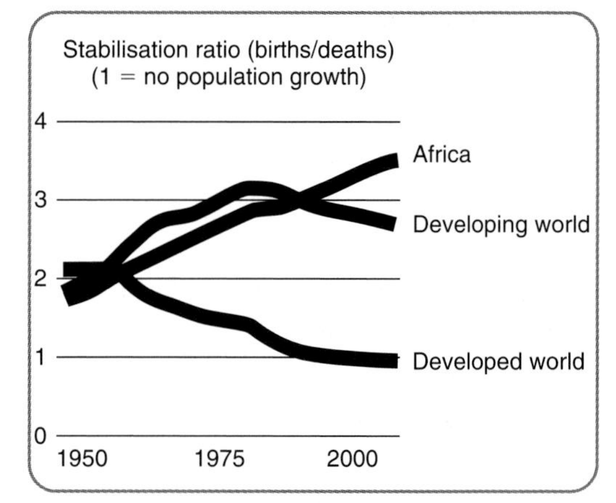

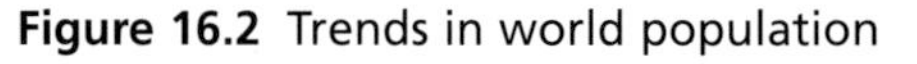
Figure 16.2 Trends in world population

Factors contributing to exponential growth

Exponential population growth is the result of the interaction of many factors. For example, improvements in buildings, weapons and settlement patterns have virtually eliminated predators of humans. Improved social and housing conditions, together with vaccination, medical and health care improvements have also contributed to a declining death rate. In terms of population growth the most significant factor is the reduction in **child mortality** as every extra female child who survives to reproductive age is likely to go on to be a mother and start a family tree. It is only very recently that a mother would expect more than a small proportion of the babies she bore to survive for longer than a few days or weeks.

Many improvements in agriculture have increased food production and thus carrying capacity, and today humans have ironically also become hunter-gatherers on a previously unheard-of scale, with high-powered vessels removing fish and other food from the oceans of the world far faster than nature can possibly replace them. This has also increased the carrying capacity, but can only be on the most temporary basis, and seems certain to end in not only the sudden and traumatic loss of a considerable food source, but the irreversible destruction of some of the most varied and productive ecosystems on Earth.

Female fecundity has increased due to earlier puberty and increased life expectancy. Women are likely to be fertile for almost 40 years, whereas an average of less than 20 years was the case until the last century. In addition to this, the actual number of babies produced during this period can potentially be much greater.

In the past it was common for children to continue breastfeeding for a much longer time than is now the case. Children would still suckle up to several years of age. The milk production that is stimulated by suckling causes hormonal effects which make it unlikely for a woman to conceive whilst breastfeeding. This is especially the case in women who live in areas where food supply is low. The result is a considerable delay between pregnancies. However, cultural habits and fashions have changed, and women now tend to cut short the duration of breastfeeding, or to use bottle feeding to eliminate it altogether. The result is that pregnancies can follow each other much more rapidly, with a consequent increase in birth rates.

For Practice

Write brief notes about each of the following factors which have contributed to exponential population growth: overcoming predation; increasing food availability; reduction in child mortality; increasing life expectancy; increases in female fecundity.

Population control through birth rate reduction and the effect on population increase

Contraception, sterilisation and abortion allow humans to limit family size. The expense of bringing up children in developed countries leads to small family size, and consequently a low birth rate and population growth.

However, not all women are able – or even wish – to use contraception, sterilisation or abortion. There may be cultural and religious objections, lack of knowledge, or a lack of money or facilities. In some societies a large number of children who can work are a financial asset, or even a necessity for the survival of the family. For whatever reason, a large family size leads to a high birth rate and population increase.

Population limiting factors

Food supply

In other species populations are strictly limited by food, water and disease. Each of these factors has a huge potential for limiting the human population. However, by his ingenuity and effort, man has overcome these limits, for the present.

The increased food production that humans have achieved has been the result of many factors.

- Changes in land use. The area of land under cultivation has been greatly increased by deforestation and destruction of natural ecosystems. In much of the developed world the natural native communities have been almost completely excluded, or confined to isolated reserves. The **natural succession** has been halted and replaced by agriculture. Similar changes are now taking place in the developing world, with **deforestation** to clear space for food production and provide fuel reaching alarming proportions. If such changes continue without attention to renewal and conservation, the consequences for genetic diversity, and for life on the planet, are likely to be severe.
- The impact of agricultural chemicals. The productivity of agricultural land has been improved beyond recognition by the use of fertilisers, herbicides, fungicides and pesticides. These chemicals have allowed the human species to avoid the worst effects of food shortage despite the exponential growth of the population. However, agricultural chemicals all have an environmental impact, and their effect, particularly on water pollution levels, is a constant cause for concern. Marine pollution in particular is a serious and increasing threat to important food supplies.
- Selective plant breeding and genetic engineering. High yielding and disease-resistant varieties of crop plants are the result of selective breeding, and it is likely that genetic engineering or genetic modification of crop plants may further increase crop yields.

In global terms, the human race has managed to provide ample food for the population and even today, with more than 6.5 billion mouths to be fed, there is more than enough for all. The malnutrition and death from starvation suffered by more than half of the world's people are the result of financial or transport problems rather than a worldwide shortage of food.

Water supply

The increasing world population results in an increasing need for water. Only a small proportion is for drinking; much more is required for industry and irrigation, and developed countries use far more per head than developing countries. The continuing rise in living standards in developing countries is certain to increase the demand for water, and as people need to populate more and more inhospitable places, where natural water supplies are limited or non-existent, a crisis seems certain to develop very soon. Already some of the world's greatest rivers, such as the Colorado, run dry long before they reach the sea.

Water covers about two-thirds of the Earth's surface, but most is too salty for use. Only 2.5% of the world's water is not salty, and two-thirds of that is locked up in the ice caps and glaciers. Of what is left, about 20% is in remote areas, and much of the rest arrives at the wrong time and place, as monsoons and floods. We use about 70% of the water we have in agriculture. But the World Water Council believes that by 2020 we shall need 17% more water than exists if we are to feed the world. Today, one person in five across the world has no access to safe drinking water, and one in two lacks safe sanitation. Adequate safe water is key to good health and a proper diet.

There are some ways to begin to tackle the problem: less wasteful irrigation systems; and planting less water intensive crops.

The long-term solution can only lie in addressing all the underlying issues which contribute to the water crisis, chief amongst which are climate change, deforestation, soil erosion and desertification.

Climate change is certain to alter rainfall patterns and cause unexpected drought in some parts of the world. **Deforestation**, which is the clearing of huge areas of natural woodland to exploit timber and provide space for agriculture, adds to the problem as the burning of trees and other vegetation releases yet more carbon dioxide, whilst at the same time destroying the photosynthetic organisms which could recycle it.

Soil erosion, the loss of the top layers of soil which leaves no material suitable for plant growth, is not caused only by deforestation. Extensive cultivated fields are vulnerable to wind erosion until the crops are big enough to stabilise the soil, and a similar danger exists after crops have been harvested. The large, undivided areas necessary to allow efficient use of machinery to maximise yield further contribute to the dangers of soil erosion.

Population pressure forces people to cultivate the marginal land around deserts. These are fragile ecosystems which are just able to survive on limited water when left alone. By removing wood for fuel and water for crops and animals, and by having to overgraze the marginal land in order to provide sufficient food, the entire system soon fails and becomes desert – this is the process of **desertification**. Once started, desertification is speeded up by the changes caused by deforestation and not only reduces the land available for food production but also permanently reduces the photosynthetic area of the planet.

Disease

Disease is an important population regulating mechanism in all animals. Human ingenuity, through our knowledge of methods of infection and our constantly increasing armoury of vaccines, medicines and treatments, together with education and cultural awareness, has given us control over many of the killer diseases of the past. Possibly even more important in controlling disease is the supply of clean drinking water, the hygienic handling of food, and the safe disposal of sewage.

Historically disease has been the most important human population limiting factor, especially during major epidemics. One example, the Black Death, is known to have killed at least half of the population of Europe in the fifteenth century, and a similar proportion in much of Asia. The disease is caused by a bacterium that is susceptible to several antibiotics, and can now therefore be cured easily and completely. It could be argued that it now poses considerably less danger to life than influenza.

Epidemics nowadays tend to be on a smaller scale due to medical advances and programmes of assistance from organisations such as the World Health Organisation (WHO), but there is a long way to go. Many diseases that are easily curable and are thus almost unheard of in the developed world, go untreated in the developing world due to financial, political and communication failures. Childhood diseases such as measles, diphtheria and polio still kill many millions of children in developing countries each year, despite the existence of inexpensive and effective vaccinations. The average life expectancy in Britain is 78, in Mozambique 31, and most of Africa less than 40. Slightly more than 57 million people died in 2005. Those deaths included 10.8 million children, almost all of whom (99%) lived in low- and middle-income countries. More than half of the children died from just five preventable or treatable conditions: respiratory infections, measles, diarrhoea, malaria, and HIV/AIDS.

For Practice

1 The problems of world population are complex. Copy the table below and use this book and other resources to add examples of both costs and benefits associated with food supply. Two examples have been included to get you started. The results are certain to show you how difficult the problems are.

Costs	Benefits
Deforestation causes loss of genetic diversity	Can support increasing population

2 Produce similar cost/benefit tables for water supply and disease control.

Population effects on the environment

It is a desperate irony that the very methods which have allowed us to continue feeding the exploding world population also constitute the chief threats to our future survival, and that of our entire biosphere.

Disruption of food webs

Natural ecosystems progress by a system of succession to what is called a climax community. This is the combination of animals and plants which natural selection has evolved to make best use of the limited resources in a particular ecosystem.

Succession is always associated with more and more species joining the ecosystem – a state of **biodiversity**. Climax communities are characterised by their biodiversity in which many species are interlinked into complex food webs that maximise the stability and adaptability of the system.

Climax communities are not, however, very productive in terms of providing food for people, and in the face of our exploding population we have been forced to prevent succession, and to replace it with near monocultures. This is achieved by the use of a cocktail of chemicals such as fertilisers, pesticides and herbicides which kill any species that may compete with the crop. Agricultural systems are massively more productive, but their lack of diversity means that they are unstable, and are extremely vulnerable to even quite small changes in conditions. Loss of complexity leads to instability.

Effects of chemicals on wildlife

The more complex the food web the more stable and adaptable the system. With lots of different species at every level, and with many criss-crossing pathways of energy flow, it is much more likely that any environmental change can be compensated for. The loss of one food source is not a problem if there are plenty of others to turn to. But in simple food webs the alternatives are very limited, and even minor changes can prove disastrous.

The effect of agricultural chemicals continues long after they have had their effect on the farm. Pesticides and fertilisers are washed into the waterways, polluting fresh and sea water with nitrates and phosphates. The steady flow of agricultural and industrial chemicals into the sea disrupts marine food webs. The sea provides a significant amount of human food and marine algae provide a large proportion of the world's oxygen by photosynthesis – all this is at risk.

For Practice

1 Use the words in the list to complete the summary of effects of agricultural chemicals on wildlife.

List: diversity; productivity; succession; species; complex; stable; unstable.

Agricultural chemicals increase the __________ of the land. They do this by reducing the number of ____________ in the habitat. This loss of __________ leads to simple foodwebs which are ___________ and cannot adapt to changing environmental conditions. Natural plant _________ increases the number of species and leads to __________ food webs which are much more __________.

Disruption of the nitrogen cycle

The term **nitrogen cycle** is used to describe the way in which proteins in living organisms are constantly being recycled. The key element in proteins is nitrogen, which is not found in either carbohydrates or fats. Nitrogen exists in a variety of forms and the importance of the nitrogen cycle is in the reprocessing of proteins. Figure 16.3 should allow you to see some of the stages that are susceptible to disruption by human activity.

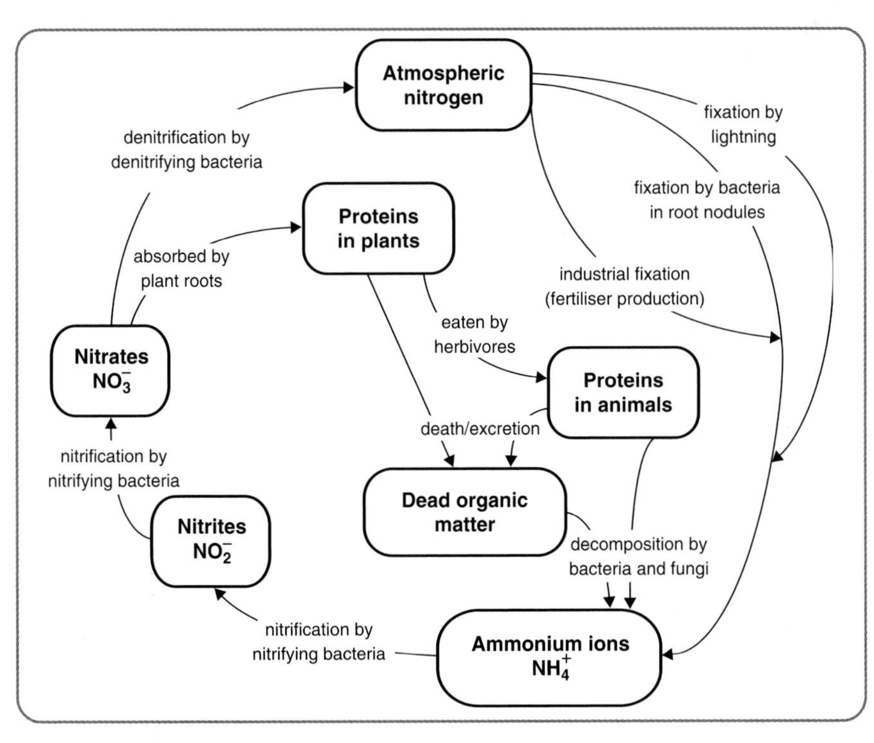

Figure 16.3 Nitrogen cycle

The productivity required to feed the increasing population needs the addition of artificial fertilisers. These are produced mostly from atmospheric nitrogen by the use of massive amounts of energy derived from fossil fuels, but dung and slurry from industrial farming is also used. Much of the applied fertiliser is washed through the soil into water courses and eventually into the sea. The rest ends up in the crops and must sooner or later decay and add yet more nitrogen compounds to aquatic ecosystems. Inadequate sewage treatment further increases the release of nitrites and nitrates into aquatic ecosystems. Sewage disposal is a serious problem with an expanding population. Solutions to this problem include more efficient treatment of sewage and methods of the removal of nitrites and nitrates. These extra nitrogen sources stimulate the growth of aquatic plants, particularly algae, causing **algal blooms** which in some cases are toxic to animals, and even if they are non-toxic, once the algae die they form a rich food supply for bacteria which multiply rapidly, depleting the oxygen in the water and resulting in damage or extinction of many animal species.

Nitrates in drinking water seem to be relatively harmless to adults, but infants convert nitrates to nitrites in their digestive systems, and nitrites interfere with the activity and production of haemoglobin. This in turn can lead to oxygen shortage in the tissues, with potentially harmful consequences for physical and mental development. High nitrite levels have been associated with **blue baby syndrome** in which infants have a blue-grey tinge to the skin caused by chronic oxygen deficiency.

Disruption of the carbon cycle

Burning fossil fuels releases massive amounts of carbon dioxide into the atmosphere, which was removed from it by photosynthesis millions of years ago. Without this imbalance the processes of photosynthesis and respiration proceed in tandem, and maintain a balance in the atmosphere. Respiration uses oxygen to release energy from food and converts the food to carbon dioxide and water. Photosynthesis removes carbon dioxide from the atmosphere and uses solar energy to combine it with water to form food and oxygen. This is the **carbon cycle**, which is summarised in Figure 16.4.

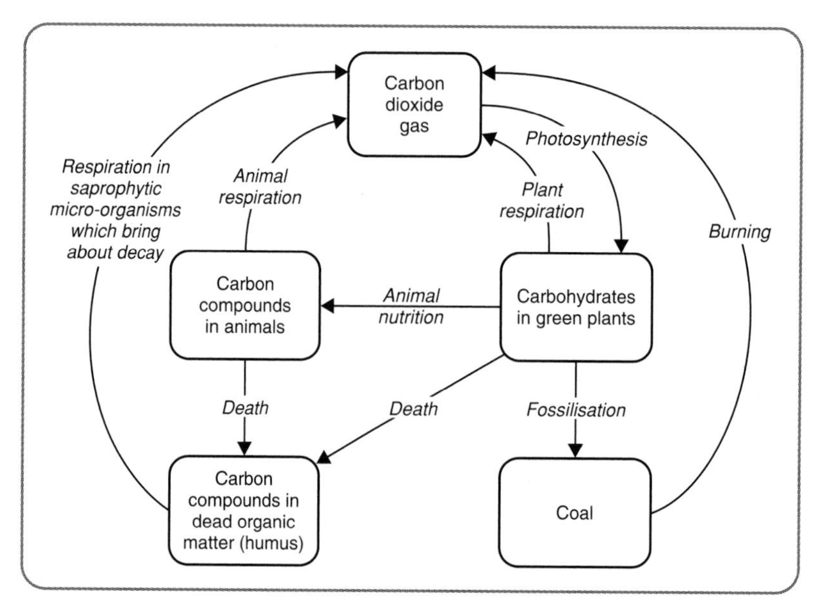

Figure 16.4 The carbon cycle

However, the removal of forests, the pollution of oceans and the covering of large areas of the planet with concrete roads and buildings, combined with the massive expansion of desert areas due to water removal and soil erosion resulting from loss of natural vegetation, is steadily reducing the photosynthetic capacity of the Earth. This loss, together with the huge release of carbon dioxide and other gases from fossil fuels, has led to a breakdown of this balance, and the concentration of carbon dioxide in our atmosphere is steadily increasing. The greenhouse effect, caused largely by carbon dioxide but increased by other gases, is causing global warming and changing weather patterns.

As Figure 16.4 shows, there are many routes by which carbon dioxide is released to the atmosphere, but only photosynthesis is capable of removing significant amounts.

The nature of ecological imbalance is that many factors interact, and any disturbance will influence many aspects of the ecosystem. In particular, the increase in atmospheric carbon dioxide caused by increasing use of fossil fuels and loss of photosynthetic capacity is giving cause for grave concern because of the **greenhouse effect** and the climate changes which are the result. The build-up of carbon dioxide in the atmosphere causes heat to be trapped instead of radiating into outer space. This is the greenhouse effect.

Methane, produced by rotting waste, cattle and paddy fields, is another 'greenhouse gas'. Increasing population is certain to lead to more household waste, cattle and paddy fields and thus more methane. These atmospheric changes are increasing average temperatures by several degrees Celsius (**global warming**), and unless the problem is addressed quickly, it is predicted that sea levels will rise due polar ice melting and low-lying areas will be extensively and frequently flooded. This will mean a loss of valuable land that is now used for homes and agriculture. Global warming is also expected to alter the direction of winds and currents and thus weather and rainfall patterns. The worldwide effects of El Niño, an occasional local disruption of ocean currents which has drastic effects on climate, winds and rainfall all over the world every 20 or so years, give us a clear example of the potential seriousness of the problem.

Rainfall patterns will be disrupted, with some areas getting more rainfall and some much less; but in either case it will not be the rainfall to which the various ecosystems are adapted. Agriculture will have to adapt crops and methods to fit the new circumstances if food supplies are to be maintained, and the pace of change is such that there is considerable doubt about whether natural ecosystems can adapt quickly enough to survive.

The future for the human race and for the fragile and beautiful planet that supports us is full of challenges and opportunities. It remains to be seen if our species can rise to these challenges. The most hopeful note is to remember that, so far in our history, human intellect and effort have allowed us to triumph over serious threats and dangers. Why should we therefore shrink before this next, and possibly greatest, challenge?

For Practice

1. Use Figure 16.4 to list the processes which release carbon dioxide into the atmosphere and identify which of them are increasing. Suggest ways in which the increase might be controlled.
2. Make a list of factors which affect the photosynthesic capacity of the Earth. Decide whether it is likely to increase or decrease, and consider ways of reversing the trend.
3. Make flash cards for all of the words that appear in **bold** type in the text of this chapter.

Exam Questions

1 The following factors affect the size of a population:

W = births

X = deaths

Y = immigration

Z = emigration.

The formula that would be used to calculate the change in the size of a population is:

A (X – W) – (Y + Z)

B (W + Y) – (X + Z)

C (W + Z) – (X + Y)

D (X + Y) – (W + Z)

2

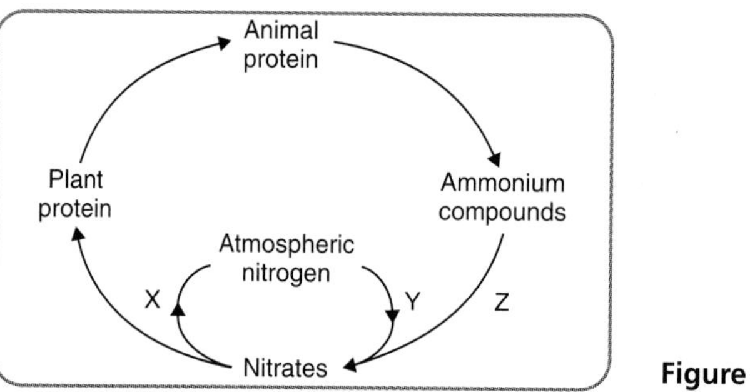

Figure 16.5

Exam Questions continued

Which line in the table identifies correctly processes X, Y and Z?

	X	Y	Z
A	Nitrogen fixation	Denitrification	Nitrification
B	Nitrification	Nitrogen fixation	Denitrification
C	Denitrification	Nitrogen fixation	Nitrification
D	Denitrification	Nitrification	Nitrogen fixation

3 Figure 16.6 shows atmospheric carbon dioxide concentrations projected up to 2100.

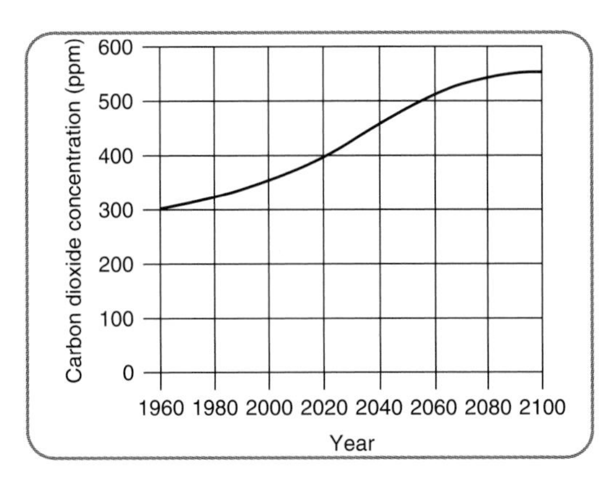

Figure 16.6 Atmospheric carbon dioxide concentrations

The percentage increase in atmospheric carbon dioxide between the years 2000 and 2100 is:

A 57

B 83

C 200

D 555.

4 Figure 16.7 outlines events that take place after fertiliser enters a loch.

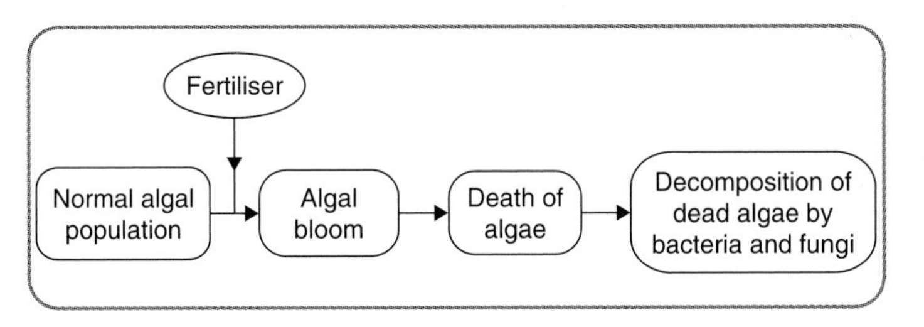

Figure 16.7 Effect of fertilisers on a loch

Exam Questions continued

(a) Name one chemical nutrient present in fertiliser that causes an algal bloom. *(1)*

(b) Explain why an algal bloom causes the death of other water plants in the loch. *(1)*

(c) Describe how the decomposition of dead algae by bacteria and fungi can lead to the death of fish in the loch. *(2)*

5 Figure 16.8 shows the population of a country as a percentage distribution by age in the year 2000 and the estimate for 2040.

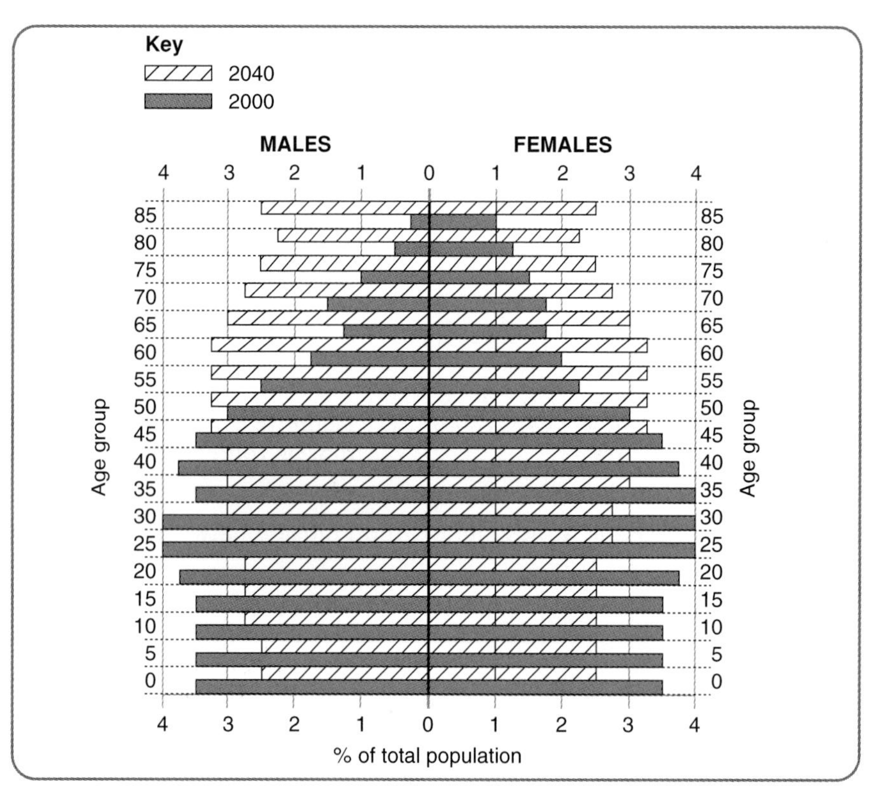

Figure 16.8 Population structure

Describe the projected changes in population between 2000 and 2040. *(2)*

Answers

1 B

2 C

3 A

4 (a) Nitrate *or* phosphate.

(b) Prevents them from getting enough light.

(c) Bacteria use the dead algae as a source of energy. Bacteria multiply. Bacteria use up all the available oxygen. (All 3 for 2 marks; any 2 for 1 mark)

5 Percentage of young decreases. Percentage of young females decreases more. Percentage of old increases. (Any two for 1 mark each)